Permaculture
21st Century Sustainability

Contents

Chapter 1

Permaculture

Permaculture is a system of agricultural and social design principles centered on simulating or directly utilizing the patterns and features observed in natural ecosystems. Permaculture was developed, and the term coined by Bill Mollison and David Holmgren in 1978.[1]

It has many branches that include but are not limited to ecological design, ecological engineering, environmental design, construction and integrated water resources management that develops sustainable architecture, and regenerative and self-maintained habitat and agricultural systems modeled from natural ecosystems.[2][3]

Mollison has said: "Permaculture is a philosophy of working with, rather than against nature; of protracted and thoughtful observation rather than protracted and thoughtless labor; and of looking at plants and animals in all their functions, rather than treating any area as a single product system."[4]

1.1 History

In 1929, Joseph Russell Smith took up an antecedent term as the subtitle for *Tree Crops: A Permanent Agriculture*, a book in which he summed up his long experience experimenting with fruits and nuts as crops for human food and animal feed.[5] Smith saw the world as an inter-related whole and suggested mixed systems of trees and crops underneath. This book inspired many individuals intent on making agriculture more sustainable, such as Toyohiko Kagawa who pioneered forest farming in Japan in the 1930s.[6]

The definition of permanent agriculture as that which can be sustained indefinitely was supported by Australian P. A. Yeomans in his 1964 book *Water for Every Farm*. Yeomans introduced an observation-based approach to land use in Australia in the 1940s, and the keyline design as a way of managing the supply and distribution of water in the 1950s.

Stewart Brand's works were an early influence noted by Holmgren.[7] Other early influences include Ruth Stout and Esther Deans, who pioneered no-dig gardening, and

Masanobu Fukuoka who, in the late 1930s in Japan, began advocating no-till orchards, gardens, and natural farming.[8]

Bill Mollison in January 2008.

In the late 1960s, Bill Mollison and David Holmgren started developing ideas about stable agricultural systems on the southern Australian island state of Tasmania. This was a result of the danger of the rapidly growing use of industrial-agricultural methods. In their view,[9] these methods were highly dependent on non-renewable resources, and were additionally poisoning land and water, reducing biodiversity, and removing billions of tons of topsoil from previously fertile landscapes. A design approach called *permaculture* was their response and was first made public with the publication of their book *Permaculture One* in 1978.[9]

By the early 1980s, the concept had broadened from agricultural systems design towards sustainable human habitats. After *Permaculture One*, Mollison further refined and developed the ideas by designing hundreds of permaculture sites and writing more detailed books, notably *Permaculture: A Designers Manual*. Mollison lectured in over 80 countries and taught his two-week Permaculture Design Course (PDC) to many hundreds of students. Mollison "encouraged graduates to become teachers themselves and set up their own institutes and demonstration sites. This multiplier effect was critical to permaculture's rapid expansion."[10]

1.2 Core tenets and principles of design

The three core tenets of permaculture are:[11][12][13]

- *Care for the earth*: Provision for all life systems to continue and multiply. This is the first principle, because without a healthy earth, humans cannot flourish.

- *Care for the people*: Provision for people to access those resources necessary for their existence.

- *Return of surplus*: Reinvesting surpluses back into the system to provide for the first two ethics. This includes returning waste back into the system to recycle into usefulness.[14] The third ethic is sometimes referred to as Fair Share to reflect that each of us should take no more than what we need before we reinvest the surplus.

Permaculture design emphasizes patterns of landscape, function, and species assemblies. It determines where these elements should be placed so they can provide maximum benefit to the local environment. The central concept of permaculture is maximizing useful connections between components and synergy of the final design. The focus of permaculture, therefore, is not on each separate element, but rather on the relationships created among elements by the way they are placed together; the whole becoming greater than the sum of its parts. Permaculture design therefore seeks to minimize waste, human labor, and energy input by building systems with maximal benefits between design elements to achieve a high level of synergy. Permaculture designs evolve over time by taking into account these relationships and elements and can become extremely complex systems that produce a high density of food and materials with minimal input.[15]

The design principles which are the conceptual foundation of permaculture were derived from the science of systems ecology and study of pre-industrial examples of sustainable land use. Permaculture draws from several disciplines including organic farming, agroforestry, integrated farming, sustainable development, and applied ecology.[16] Permaculture has been applied most commonly to the design of housing and landscaping, integrating techniques such as agroforestry, natural building, and rainwater harvesting within the context of permaculture design principles and theory.

1.3 Theory

1.3.1 Twelve design principles

Twelve Permaculture design principles articulated by David Holmgren in his *Permaculture: Principles and Pathways Beyond Sustainability*:[17]

1. *Observe and interact*: By taking time to engage with nature we can design solutions that suit our particular situation.

2. *Catch and store energy*: By developing systems that collect resources at peak abundance, we can use them in times of need.

3. *Obtain a yield*: Ensure that you are getting truly useful rewards as part of the work that you are doing.

4. *Apply self-regulation and accept feedback*: We need to discourage inappropriate activity to ensure that systems can continue to function well.

5. *Use and value renewable resources and services*: Make the best use of nature's abundance to reduce our consumptive behavior and dependence on non-renewable resources.

6. *Produce no waste*: By valuing and making use of all the resources that are available to us, nothing goes to waste.

7. *Design from patterns to details*: By stepping back, we can observe patterns in nature and society. These can form the backbone of our designs, with the details filled in as we go.

8. *Integrate rather than segregate*: By putting the right things in the right place, relationships develop between those things and they work together to support each other.

9. *Use small and slow solutions*: Small and slow systems are easier to maintain than big ones, making better use of local resources and producing more sustainable outcomes.

10. *Use and value diversity*: Diversity reduces vulnerability to a variety of threats and takes advantage of the unique nature of the environment in which it resides.

11. *Use edges and value the marginal*: The interface between things is where the most interesting events take place. These are often the most valuable, diverse and productive elements in the system.

12. *Creatively use and respond to change*: We can have a positive impact on inevitable change by carefully observing, and then intervening at the right time.

1.3.2 Layers

Suburban permaculture garden in Sheffield, UK with different layers of vegetation

Layers are one of the tools used to design functional ecosystems that are both sustainable and of direct benefit to humans. A mature ecosystem has a huge number of relationships between its component parts: trees, understory, ground cover, soil, fungi, insects, and animals. Because plants grow to different heights, a diverse community of life is able to grow in a relatively small space, as the vegetation occupies different layers. There are generally seven recognized layers in a food forest, although some practitioners also include fungi as an eighth layer.[18]

1. The canopy: the tallest trees in the system. Large trees dominate but typically do not saturate the area, i.e. there exist patches barren of trees.

2. Understory layer: trees that revel in the dappled light under the canopy.

3. Shrub layer: a diverse layer of woody perennials of limited height. includes most berry bushes.

4. Herbaceous layer: Plants in this layer die back to the ground every winter (if winters are cold enough, that is). They do not produce woody stems as the Shrub layer does. Many culinary and medicinal herbs are in this layer. A large variety of beneficial plants fall into this layer. May be annuals, biennials or perennials.

5. Soil surface/Groundcover: There is some overlap with the Herbaceous layer and the Groundcover layer; however plants in this layer grow much closer to the ground, grow densely to fill bare patches of soil, and often can tolerate some foot traffic. Cover crops retain soil and lessen erosion, along with green manures that add nutrients and organic matter to the soil, especially nitrogen.

6. Rhizosphere: Root layers within the soil. The major components of this layer are the soil and the organisms that live within it such as plant roots (including root crops such as potatoes and other edible tubers), fungi, insects, nematodes, worms, etc.

7. Vertical layer: climbers or vines, such as runner beans and lima beans (vine varieties).[18][19]

1.3.3 Guilds

There are many forms of guilds, including guilds of plants with similar functions (that could interchange within an ecosystem), but the most common perception is that of a mutual support guild. Such a guild is a group of species where each provides a unique set of diverse functions that work in conjunction, or harmony. Mutual support guilds are groups of plants, animals, insects, etc. that work well together. Some plants may be grown for food production, some have tap roots that draw nutrients up from deep in the soil, some are nitrogen-fixing legumes, some attract beneficial insects, and others repel harmful insects. When grouped together in a mutually beneficial arrangement, these plants form a guild. See Dave Jacke's work on edible forest gardens for more information on other guilds, specifically resource-partitioning and community-function guilds.[20][21][22]

1.3.4 Edge effect

The edge effect in ecology is the effect of the juxtaposition or placing side by side of contrasting environments on an ecosystem. Permaculturists argue that, where vastly differing systems meet, there is an intense area of productivity and useful connections. An example of this is the coast; where the land and the sea meet there is a particularly rich area that meets a disproportionate percentage of human and animal needs. So this idea is played out in permacultural designs by using spirals in the herb garden or creating ponds that have wavy undulating shorelines rather than a simple circle or oval (thereby increasing the amount of edge for a given area).

1.3.5 Zones

Zones are a way of intelligently organizing design elements in a human environment on the basis of the frequency of human use and plant or animal needs. Frequently manipulated or harvested elements of the design are located close to the house in zones 1 and 2. Less frequently used or manipulated elements, and elements that benefit from isolation

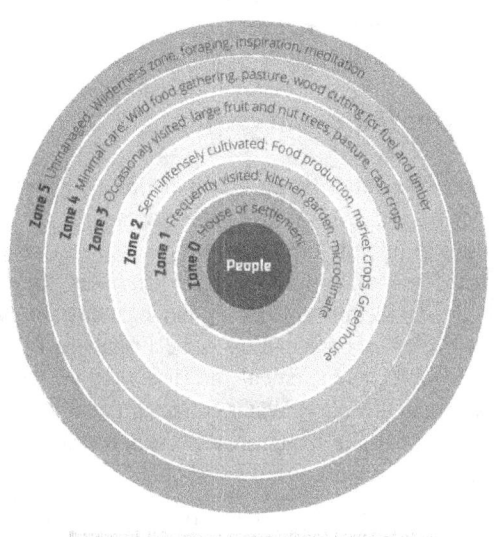

Permaculture Zones 0-5.

(such as wild species) are farther away. Zones are about positioning things appropriately, and are numbered from 0 to 5.[23]

Zone 0 The house, or home center. Here permaculture principles would be applied in terms of aiming to reduce energy and water needs, harnessing natural resources such as sunlight, and generally creating a harmonious, sustainable environment in which to live and work. Zone 0 is an informal designation, which is not specifically defined in Bill Mollison's book.

Zone 1 The zone nearest to the house, the location for those elements in the system that require frequent attention, or that need to be visited often, such as salad crops, herb plants, soft fruit like strawberries or raspberries, greenhouse and cold frames, propagation area, worm compost bin for kitchen waste, etc. Raised beds are often used in zone 1 in urban areas.

Zone 2 This area is used for siting perennial plants that require less frequent maintenance, such as occasional weed control or pruning, including currant bushes and orchards, pumpkins, sweet potato, etc. This would also be a good place for beehives, larger scale composting bins, etc.

Zone 3 The area where main-crops are grown, both for domestic use and for trade purposes. After establishment, care and maintenance required are fairly minimal (provided mulches and similar things are used), such as watering or weed control maybe once a week.

Zone 4 A semi-wild area. This zone is mainly used for forage and collecting wild food as well as production of timber for construction or firewood.

Zone 5 A wilderness area. There is no human intervention in zone 5 apart from the observation of natural ecosystems and cycles. Through this zone we build up a natural reserve of bacteria, moulds and insects that can aid the zones above it.[24]

1.3.6   People and permaculture

Permaculture uses observation of nature to create regenerative systems, and the place where this has been most visible has been on the landscape. There has been a growing awareness though that firstly, there is the need to pay more attention to the peoplecare ethic, as it is often the dynamics of people that can interfere with projects, and secondly that the principles of permaculture can be used as effectively to create vibrant, healthy and productive people and communities as they have been in landscapes.

1.3.7   Domesticated animals

Domesticated animals are often incorporated into site design.[25]

1.4   Common practices

1.4.1   Agroforestry

Agroforestry is an integrated approach of using the interactive benefits from combining trees and shrubs with crops and/or livestock. It combines agricultural and forestry technologies to create more diverse, productive, profitable, healthy and sustainable land-use systems.[26] In agroforestry systems, trees or shrubs are intentionally used within agricultural systems, or non-timber forest products are cultured in forest settings.

Forest gardening is a term permaculturalists use to describe systems designed to mimic natural forests. Forest gardens, like other permaculture designs, incorporate processes and relationships that the designers understand to be valuable in natural ecosystems. The terms forest garden and food forest are used interchangeably in the permaculture literature. Numerous permaculturists are proponents of forest gardens, such as Graham Bell, Patrick Whitefield, Dave Jacke, Eric Toensmeier and Geoff Lawton. Bell started building his forest garden in 1991 and wrote the book *The Permaculture Garden* in 1995, Whitefield wrote the book

How to Make a Forest Garden in 2002, Jacke and Toensmeier co-authored the two volume book set *Edible Forest Gardening* in 2005, and Lawton presented the film *Establishing a Food Forest* in 2008.[15][27][28]

Tree Gardens, such as Kandyan tree gardens, in South and Southeast Asia, are often hundreds of years old. Whether they derived initially from experiences of cultivation and forestry, as is the case in agroforestry, or whether they derived from an understanding of forest ecosystems, as is the case for permaculture systems, is not self-evident. Many studies of these systems, especially those that predate the term permaculture, consider these systems to be forms of agroforestry. Permaculturalists who include existing and ancient systems of polycropping with woody species as examples of food forests may obscure the distinction between permaculture and agroforestry.

Food forests and agroforestry are parallel approaches that sometimes lead to similar designs.

1.4.2 Hügelkultur

Hügelkultur is the practice of burying large volumes of wood to increase soil water retention. The porous structure of wood acts as a sponge when decomposing underground. During the rainy season, masses of buried wood can absorb enough water to sustain crops through the dry season.[29] This technique has been used by permaculturalists Sepp Holzer, Toby Hemenway, Paul Wheaton, and Masanobu Fukuoka.[30][31]

1.4.3 Natural building

A natural building involves a range of building systems and materials that place major emphasis on sustainability. Ways of achieving sustainability through natural building focus on durability and the use of minimally processed, plentiful or renewable resources, as well as those that, while recycled or salvaged, produce healthy living environments and maintain indoor air quality.

The basis of natural building is the need to lessen the environmental impact of buildings and other supporting systems, without sacrificing comfort, health, or aesthetics. To be more sustainable, natural building uses primarily abundantly available, renewable, reused, or recycled materials. In addition to relying on natural building materials, the emphasis on the architectural design is heightened. The orientation of a building, the utilization of local climate and site conditions, the emphasis on natural ventilation through design, fundamentally lessen operational costs and positively impact the environment. Building compactly and minimizing the ecological footprint is common, as are on-site handling of energy acquisition, on-site water capture, alternate sewage treatment, and water reuse.

1.4.4 Rainwater harvesting

Rainwater harvesting is the accumulating and storing of rainwater for reuse before it reaches the aquifer.[32] It has been used to provide drinking water, water for livestock, water for irrigation, as well as other typical uses. Rainwater collected from the roofs of houses and local institutions can make an important contribution to the availability of drinking water. It can supplement the subsoil water level and increase urban greenery. Water collected from the ground, sometimes from areas which are especially prepared for this purpose, is called stormwater harvesting.

Greywater is wastewater generated from domestic activities such as laundry, dishwashing, and bathing, which can be recycled on-site for uses such as landscape irrigation and constructed wetlands. Greywater is largely sterile, but not potable (drinkable). Greywater differs from water from the toilets, which is designated sewage or blackwater to indicate it contains human waste. Blackwater is septic or otherwise toxic and cannot easily be reused. There are, however, continuing efforts to make use of blackwater or human waste. The most notable is for composting through a process known as humanure; a combination of the words human and manure. Additionally, the methane in humanure can be collected and used similar to natural gas as a fuel, such as for heating or cooking, and is commonly referred to as biogas. Biogas can be harvested from the human waste and the remainder still used as humanure. Some of the simplest forms of humanure use include a composting toilet or an outhouse or dry bog surrounded by trees that are heavy feeders which can be coppiced for wood fuel. This process eliminates the use of a standard toilet with plumbing.

1.4.5 Sheet mulching

In agriculture and gardening, mulch is a protective cover placed over the soil. Any material or combination can be used as mulch, such as stones, leaves, cardboard, wood chips, gravel, etc., though in permaculture mulches of organic material are the most common because they perform more functions. These include absorbing rainfall, reducing evaporation, providing nutrients, increasing organic matter in the soil, feeding and creating habitat for soil organisms, suppressing weed growth and seed germination, moderating diurnal temperature swings, protecting against frost, and reducing erosion. Sheet mulching is an agricultural no-dig gardening technique that attempts to mimic natural processes occurring within forests. Sheet mulching mimics the leaf cover that is found on forest floors. When deployed properly and in combination with other Permacultural principles, it can generate healthy, productive and low maintenance ecosystems.[33][34]

Sheet mulch serves as a "nutrient bank," storing the nutrients contained in organic matter and slowly making these nutrients available to plants as the organic matter slowly and naturally breaks down. It also improves the soil by attracting and feeding earthworms, slaters and many other soil microorganisms, as well as adding humus. Earthworms "till" the soil, and their worm castings are among the best fertilizers and soil conditioners. Sheet mulching can be used to reduce or eliminate undesirable plants by starving them of light, and can be more advantageous than using herbicide or other methods of control.

1.4.6 Intensive rotational grazing

Grazing has long been blamed for much of the destruction we see in the environment. However, it has been shown that when grazing is modeled after nature, the opposite effect can be seen.[35][36] Also known as cell grazing, managed intensive rotational grazing (MIRG) is a system of grazing in which ruminant and non-ruminant herds and/or flocks are regularly and systematically moved to fresh pasture, range, or forest with the intent to maximize the quality and quantity of forage growth. This disturbance is then followed by a period of rest which allows new growth. MIRG can be used with cattle, sheep, goats, pigs, chickens, rabbits, geese, turkeys, ducks, and other animals depending on the natural ecological community that is being mimicked. Sepp Holzer and Joel Salatin have shown how the disturbance caused by the animals can be the spark needed to start ecological succession or prepare ground for planting. Allan Savory's holistic management technique has been likened to "a permaculture approach to rangeland management".[37][38] One variation on MIRG that is gaining rapid popularity is called eco-grazing. Often used to either control invasives or re-establish native species, in eco-grazing the primary purpose of the animals is to benefit the environment and the animals can be, but are not necessarily, used for meat, milk or fiber.[39][40][41][42][43][44][45]

1.4.7 Keyline design

Keyline design is a technique for maximizing beneficial use of water resources of a piece of land developed in Australia by farmer and engineer P. A. Yeomans. The *Keyline* refers to a specific topographic feature linked to water flow which is used in designing the drainage system of the site.[46]

1.4.8 Fruit tree management

Some proponents of permaculture advocate no, or limited, pruning. One advocate of this approach is Sepp Holzer who used the method in connection with Hügelkultur berms. He has successfully grown several varieties of fruiting trees at altitudes (approximately 9,000 feet (2,700 m)) far above their normal altitude, temperature, and snow load ranges. He notes that the Hügelkultur berms kept and/or generated enough heat to allow the roots to survive during alpine winter conditions. The point of having unpruned branches, he notes, was that the longer (more naturally formed) branches bend over under the snow load until they touched the ground, thus forming a natural arch against snow loads that would break a shorter, pruned, branch.

Masanobu Fukuoka, as part of early experiments on his family farm in Japan, experimented with no-pruning methods, noting that he ended up killing many fruit trees by simply letting them go, which made them become convoluted and tangled, and thus unhealthy.[47][48] Then he realised this is the difference between natural-form fruit trees and the process of change of tree form that results from abandoning previously-pruned unnatural fruit trees.[47][49] He concluded that the trees should be raised all their lives without pruning, so they form healthy and efficient branch patterns that follow their natural inclination. This is part of his implementation of the Tao-philosophy of Wú wéi translated in part as no-action (against nature), and he described it as no unnecessary pruning, nature farming or "do-nothing" farming, of fruit trees, distinct from non-intervention or literal no-pruning. He ultimately achieved yields comparable to or exceeding standard/intensive practices of using pruning and chemical fertilisation.[47][49][50]

1.5 Trademark and copyright issues

There has been contention over who, if anyone, controls legal rights to the word *permaculture*: is it trademarked or copyrighted? and if so, who holds the legal rights to the use of the word? For a long time Bill Mollison claimed to have copyrighted the word, and his books said on the copyright page, "The contents of this book and the word PERMACULTURE are copyright." These statements were largely accepted at face-value within the permaculture community. However, copyright law does not protect names, ideas, concepts, systems, or methods of doing something; it only protects the expression or the description of an idea, not the idea itself. Eventually Mollison acknowledged that he was mistaken and that no copyright protection existed for the word *permaculture*.[51]

In 2000, Mollison's US based Permaculture Institute sought a service mark (a form of trademark) for the word *per-

maculture when used in educational services such as conducting classes, seminars, or workshops.[52] The service mark would have allowed Mollison and his two Permaculture Institutes (one in the US and one in Australia) to set enforceable guidelines regarding how permaculture could be taught and who could teach it, particularly with relation to the PDC, despite the fact that he had instituted a system of certification of teachers to teach the PDC in 1993. The service mark failed and was abandoned in 2001. Also in 2001 Mollison applied for trademarks in Australia for the terms "Permaculture Design Course"[53] and "Permaculture Design".[53] These applications were both withdrawn in 2003. In 2009 he sought a trademark for "Permaculture: A Designers' Manual"[53] and "Introduction to Permaculture",[53] the names of two of his books. These applications were withdrawn in 2011. There has never been a trademark for the word *permaculture* in Australia.[53]

1.6 Criticisms

1.6.1 General criticisms

In 2011, Owen Hablutzel argued that "permaculture has yet to gain a large amount of specific mainstream scientific acceptance," and that "the sensitiveness to being perceived and accepted on scientific terms is motivated in part by a desire for permaculture to expand and become increasingly relevant."

In his books *Sustainable Freshwater Aquaculture* and *Farming in Ponds and Dams*, Nick Romanowski expresses the view that the presentation of aquaculture in Bill Mollison's books is unrealistic and misleading.[54]

1.6.2 Agroforestry

Greg Williams argues that forests cannot be more productive than farmland because the net productivity of forests decline as they mature due to ecological succession.[55] Proponents of permaculture respond that this is true only if one compares data between woodland forest and climax vegetation, but not when comparing farmland vegetation with woodland forest. For example, ecological succession generally results in a forest's productivity rising after its establishment only until it reaches the *woodland state* (67% tree cover), before declining until *full maturity*.[15]

1.7 See also

- Agrarianism

- Agroecology

- Agroforestry

- Aquaponics

- Biodynamics

- Bill Mollison

- Biointensive agriculture

- Biomimicry

- Climate-friendly gardening

- David Holmgren

- Ecoagriculture

- Food-feed system

- Forest gardening

- Geoff Lawton

- Holzer Permaculture

- Hügelkultur

- List of permaculture projects

- Microponics

- Paul Wheaton

- Permaforestry

- Regenerative agriculture

- Seed saving

- Sepp Holzer

- Zaï

1.8 References

[1] Introduction to Permaculture, (1991), Mollison, p. v

[2] Hemenway 2009, p. 5.

[3] Mars, Ross (2005). *The Basics of Permaculture Design*. Chelsea Green. p. 1. ISBN 978-1-85623-023-0.

[4] Mollison, B. (1991). *Introduction to permaculture*. Tasmania, Australia: Tagari.

[5] Smith, Joseph Russell; Smith, John (1987). *Tree Crops: A permanent agriculture*. Island Press. ISBN 978-1-59726873-8.

[6] Hart 1996, p. 41.

[7] Holmgren, David (2006). "The Essence of Permaculture". Holmgren Design Services. Retrieved 10 September 2011.

[8] Mollison, Bill (September 15–21, 1978). "The One-Straw Revolution by Masanobu Fukuoka". *Nation Review*. p. 18.

[9] *Introduction to Permaculture*, 1991, Mollison, p.v

[10] Lillington, Ian; Holmgren, David; Francis, Robyn; Rosenfeldt, Robyn. "The Permaculture Story: From 'Rugged Individuals' to a Million Member Movement" (PDF). *Pip Magazine*. Retrieved 9 July 2015.

[11] Greenblott, Kara; Nordin, Kristof (2012), *Permaculture Design for Orphans and Vulnerable Children Programming: Low-Cost, Sustainable Solutions for Food and Nutrition Insecure Communities*, AIDS Support and Technical Assistance Resources, AIDSTAR-One (Task Order 1), Arlington, VA: USAID.

[12] Mollison 1988, p. 2.

[13] Holmgren, David (2002). *Permaculture: Principles & Pathways Beyond Sustainability*. Holmgren Design Services. p. 1. ISBN 0-646-41844-0.

[14] Mollison, Bill. "Permaculture: A Quiet Revolution". *Scott London* (interview). Retrieved 17 May 2013.

[15] "Edible Forest Gardening".

[16] Holmgren, David (1997). "Weeds or Wild Nature" (PDF). Permaculture International Journal. Retrieved 10 September 2011.

[17] "Permaculture: Principles and Pathways Beyond Sustainability". Holmgren Design. Retrieved 2013-10-21.

[18] *Nine layers of the edible forest garden*, TC permaculture, May 27, 2013.

[19] "Seven layers of a forest", *Food forests*, CA: Permaculture school.

[20] Simberloff, D; Dayan, T (1991). "The Guild Concept and the Structure of Ecological Communities". *Annual Review of Ecology and Systematics*. **22**: 115. doi:10.1146/annurev.es.22.110191.000555.

[21] "Guilds". *Encyclopaedia Britannica*. Retrieved 2011-10-21.

[22] Williams, SE; Hero, JM (1998). "Rainforest frogs of the Australian Wet Tropics: guild classification and the ecological similarity of declining species". *Proceedings. Biological sciences*. The Royal Society. **265** (1396): 597–602. doi:10.1098/rspb.1998.0336. PMC 1689015. PMID 9881468.

[23] Burnett 2001.

[24] *Permacultuur course*, NL: WUR.

[25] Mollison 1988, p. 5: 'Deer, rabbits, sheep, and herbivorous fish are very useful to us, in that they convert unusable herbage to acceptable human food. Animals represent a valid method of storing inedible vegetation as food.'

[26] "USDA National Agroforestry Center (NAC)". UNL. 2011-08-01. Retrieved 2011-10-21.

[27] "Graham Bell's Forest Garden". *Permaculture*. Media mice.

[28] "Establishing a Food Forest" (film review). Transition culture. Feb 11, 2009.

[29] Wheaton, Paul. "Raised garden beds: hugelkultur instead of irrigation" Richsoil. Retrieved 2012-07-15.

[30] Hemenway 2009, pp. 84–85.

[31] Feineigle, Mark. "Hugelkultur: Composting Whole Trees With Ease". Permaculture Research Institute of Australia. Retrieved 2012-07-15.

[32] "Rainwater harvesting". DE: Aramo. 2012. Retrieved 19 August 2015.

[33] "Sheet Mulching: Greater Plant and Soil Health for Less Work". Agroforestry. 2011-09-03. Retrieved 2011-10-21.

[34] Mason, J (2003), *Sustainable Agriculture*, Landlinks.

[35] "Prince Charles sends a message to IUCN's World Conservation Congress". *International Union for Conservation of Nature*. Retrieved 6 April 2013.

[36] Undersander, Dan; et al. "Grassland birds: Fostering habitat using rotational grazing" (PDF). University of Wisconsin-Extension. Retrieved 5 April 2013.

[37] Fairlie, Simon (2010). *Meat: A Benign Extravagance*. Chelsea Green. pp. 191–93. ISBN 978-1-60358325-1.

[38] Bradley, Kirsten. "Holistic Management: Herbivores, Hats, and Hope". Milkwood. Retrieved 25 March 2014.

[39] "Munching sheep replace lawn mowers in Paris". *The Sunday Times*. Apr 4, 2013. Retrieved 7 April 2013.

[40] Ash, Andrew, *The Ecograze Project – developing guidelines to better manage grazing country* (PDF), et al., CSIRO, ISBN 0-9579842-0-0, retrieved 7 April 2013

[41] McCarthy, Caroline. "Things to make you happy: Google employs goats". *CNET*. Retrieved 7 April 2013.

[42] Gordon, Ian. "A systems approach to livestock/resource interactions in tropical pasture systems" (PDF). *The James Hutton Institute*. Retrieved 7 April 2013.

[43] Littman, Margaret. "Getting your goat: Eco-friendly mowers". *Chicago Tribune News*. Retrieved 7 April 2013.

[44] Stevens, Alexis. "Kudzu-eating sheep take a bite out of weeds". *The Atlanta Journal-Constitution*. Retrieved 7 April 2013.

[45] Klynstra, Elizabeth. "Hungry sheep invade Candler Park". *CBS Atlanta*. Retrieved 7 April 2013.

[46] Tipping, Don (4 January 2013). "Creating Permaculture Keyline Water Systems" (video). UK: Beaver State Permaculture.

[47] Masanobu, Fukuoka (1987) [1985], *The Natural Way of Farming – The Theory and Practice of Green Philosophy* (rev ed.), Tokyo: Japan Publications, p. 204

[48] Fukuoka 1978, pp. 13, 15–18, 46, 58–60.

[49] Fukuoka 1978.

[50] "Masanobu Fukuoka", *Public Service* (biography), PH: The Ramon Magsaysay Award Foundation, 1988.

[51] Grayson, Russ (2011). "The Permaculture Papers 5: time of change and challenge — 2000-2004". Pacific edge. Retrieved 8 September 2011.

[52] United States Patent and Trademark Office (2011). "Trademark Electronic Search System (TESS)". US Department of Commerce. Retrieved 8 September 2011.

[53] "Result". IP Australia. 2011. Retrieved 8 September 2011.

[54] Nick Romanowski (2007). *Sustainable Freshwater Aquaculture: The Complete Guide from Backyard to Investor*. UNSW Press. p. 130. ISBN 978-0-86840-835-4.

[55] Williams, Greg (2001). "Gaia's Garden: A Guide to Home-Scale Permaculture". *Whole Earth*.

1.8.1 Bibliography

- Bell, Graham (2004) [1992, Thorsons, ISBN 0-7225-2568-0], *The Permaculture Way* (2nd ed.), UK: Permanent Publications, ISBN 1-85623-028-7.

- —— (2004), *The Permaculture Garden*, UK: Permanent, ISBN 1-85623-027-9.

- Burnett, G (2001), *Permaculture: a Beginner's Guide*, UK: Spiralseed, ISBN 978-0-95534921-8.

- Fern, Ken (1997), *Plants For A Future*, UK: Permanent, ISBN 1-85623-011-2.

- Fukuoka, Masanobu (1978), *The One–Straw Revolution*, Holistic Agriculture Library, US: Rodale Books.

- Hart, Robert (1996), *Forest Gardening*, UK: Green Books, p. 41, ISBN 978-1-60358050-2; ISBN 1-900322-02-1.

- Hemenway, Toby (2009) [2001, ISBN 1-890132-52-7], *Gaia's Garden: A Guide to Home-Scale Permaculture*, US: Chelsea Green, ISBN 978-1-60358-029-8

- Holmgren, David, *Melliodora (Hepburn Permaculture Gardens): A Case Study in Cool Climate Permaculture 1985–2005*, AU: Holmgren Design Services.

- ——, *Collected Writings & Presentations 1978–2006*, AU: Holmgren Design Services.

- —— (2009), *Future Scenarios*, White River Junction: Chelsea Green.

- ——, *Permaculture: Principles and Pathways Beyond Sustainability*, AU: Holmgren Design Services.

- ——, *Update 49: Retrofitting the suburbs for sustainability*, AU: CSIRO Sustainability Network.

- Jacke, Dave with Eric Toensmeier. *Edible Forest Gardens. Volume I: Ecological Vision and Theory for Temperate-Climate Permaculture, Volume II: Ecological Design and Practice for Temperate-Climate Permaculture*. Edible Forest Gardens (US) 2005

- King, Franklin Hiram (1911), *Farmers of Forty Centuries: Or Permanent Agriculture in China, Korea and Japan*.

- Law, Ben (2005), *The Woodland House*, UK: Permanent, ISBN 1-85623-031-7.

- ——, *The Woodland Way*, UK: Permanent Publications, ISBN 1-85623-009-0.

- Loofs, Mona. *Permaculture, Ecology and Agriculture: An investigation into Permaculture theory and practice using two case studies in northern New South Wales* Honours thesis, Human Ecology Program, Department of Geography, Australian National University 1993

- Macnamara, Looby. *People and Permaculture: caring and designing for ourselves, each other and the planet.* [Permanent Publications] (UK) (2012) ISBN 1-85623-087-2.

- Mollison, Bill (1979), *Permaculture Two*, Australia: Tagari Press, ISBN 0-908228-00-7.

- —— (1988), *Permaculture: A Designer's Manual*, AU: Tagari Press, ISBN 0-908228-01-5.

- ——; Holmgren, David (1978), *Permaculture One*, AU: Transworld Publishers, ISBN 0-552-98060-9.

- Odum, H.T., Jorgensen, S.E. and Brown, M.T. 'Energy hierarchy and transformity in the universe', in *Ecological Modelling*, 178, pp. 17–28 (2004).

- Paull, J. "Permanent Agriculture: Precursor to Organic Farming", Journal of Bio-Dynamics Tasmania, no.83, pp. 19–21, 2006. Organic eprints.

- Rosemary, Morrow, *Earth User's Guide to Permaculture*, ISBN 0-86417-514-0.

- Shepard, Mark: *Restoration Agriculture – Redesigning Agriculture in Nature's Image*, Acres US, 2013, ISBN 1-60173035-7

- Whitefield, Patrick (1993), *Permaculture In A Nutshell*, UK: Permanent, ISBN 1-85623-003-1.

- ——— (2004), *The Earth Care Manual*, UK: Permanent Publications, ISBN 1-85623-021-X.

- Woodrow, Linda. *The Permaculture Home Garden*. Penguin Books (Australia).

- Yeomans, P.A. *Water for Every Farm: A practical irrigation plan for every Australian property*, KG Murray, Sydney, NSW, Australia (1973).

- *The Same Planet a different World* (free ebook), FR.

1.9 External links

- Ferguson, Rafter Sass; Lovell, Sarah Taylor (2013), "Permaculture for agroecology: design, movement, practice, and worldview", *Agronomy for Sustainable Development* (review), Springer, **34** (2): 251, doi:10.1007/s13593-013-0181-6 – The first systematic review of the permaculture literature, from the perspective of agroecology.

- The Permaculture Research Institute – Permaculture Forums, Courses, Information, News and Worldwide Reports.

- The Worldwide Permaculture Network – Database of permaculture people and projects worldwide.

- The Permaculture Association, UK.

- The 15 pamphlets based on the 1981 Permaculture Design Course given by Bill Mollison (co-founder of permaculture) all in 1 PDF file.

- David Holmgren's web site (co-founder of permaculture)

- Ethics and principles of permaculture (Holmgren's)

- Permaculture a Beginners Guide – a 'pictorial walkthrough'

- Permaculture – Sustainability and sustainable development

- Urban Permaculture Design – a city lot with over a hundred perennial edible varieties. Permaculture land acquisition discussion.

- A quarter acre suburban property in Eugene, Oregon – grass to garden, reclaim automobile space, elevated/edible landscape, rain water catchment, passive solar design, education

- The Permaculture Activist is a co-evolving quarterly produced by a dedicated handful of entirely part-time folks

- Permaculture Commons is a collection of permaculture material under free licenses

Chapter 2

Agroecological restoration

Agroecological restoration is the practice of re-integrating natural systems into agriculture in order to maximize sustainability, ecosystem services, and biodiversity. This is one example of a way to apply the principles of agroecology to an agricultural system.

2.1 Overview

Farms cannot be restored to a purely natural state because of the negative economic impact on farmers, but returning processes, such as pest control to nature with the method of intercropping, allows a farm to be more ecologically sustainable and, at the same time, economically viable. Agroecological restoration works toward this balance of sustainability and economic viability because conventional farming is not sustainable over the long run without the integration of natural systems and because the use of land for agriculture has been a driving force in creating the present world biodiversity crisis. Its efforts are complementary to, rather than a substitute for, biological conservation.[1]

"...biodiversity is just as important on farms and in fields as it is in deep river valleys or mountain cloud forests."

FAO, 15 October 2004

Agriculture creates a conflict over the use of land between wildlife and humans. Though the domestication of crop plants occurred 10,000 years ago, a 500% increase in the amount of pasture and crop land over the last three hundred years has led to the rapid loss of natural habitats.[2] In recent years, the world community acknowledged the value of biodiversity in treaties, such as the 1992 landmark Convention on Biological Diversity.[3]

2.2 Reintegration

The reintegration of agricultural systems into more natural systems will result in decreased yield and produce a more complex system, but there will be considerable gains in biodiversity and ecosystem services.

2.2.1 Biodiversity

The Food and Agriculture Organization of the United Nations estimates that more than 40% of earth's land surface is currently used for agriculture. And because so much land has been converted to agriculture, habitat loss is recognized as the driving force in biodiversity loss (FAO). This biodiversity loss often occurred in two steps, as in the American Midwest, with the introduction of mixed farming carried out on small farms and then with the widespread use of mechanized farming and monoculture beginning after World War II.[4] The decline in farmland biodiversity can now be traced to changes in farming practices and increased agricultural intensity.[5]

2.2.2 Increasing heterogeneity

Heterogeneity (here, the diversity or complexity of the landscape) has been shown to be associated with species diversity. For example, the abundance of butterflies has been found to increase with heterogeneity. One important part of maintaining heterogeneity in the spaces between different fields is made up of habitat that is not cropped, such as grass margins and strips, scrub along field boundaries, woodland, ponds, and fallow land. These seemingly unimportant pieces of land are crucial for the biodiversity of a farm. The presence of field margins benefits many different taxa: the plants attract herbivorous insects, will which attract certain species of birds and those birds will attract their natural predators. Also, the cover provided by the no cropped habitat allows the species that need a large range to move across the landscape.[6]

2.2.3 Monoculture

In the absence of cover, species face a landscape in which their habitat is greatly fragmented. The isolation of a species to a small habitat that it can't safely wander from can create a genetic bottleneck, decreasing the resilience of the particular population, and be another factor leading to the decline of the total population of the species.[7] Monoculture, the practice of producing a single crop over a wide area, causes fragmentation. In conventional farming, monoculture, such as with rotations of corn and soybean crops planted in alternating growing seasons, is used so that very high yields can be produced. After the mechanization of farming, monoculture became a standard practice in corn-beans rotation, and had broad implications for the long-term sustainability and biodiversity of farms. Whereas organic fertilizers, had kept the soil's nutrients fixed to the ecosystem, the introduction of monoculture removed the nutrients and farmers compensated for that loss by using inorganic fertilizers. It is estimated that humans have doubled the rate of nitrogen input into the nitrogen cycle, mostly since 1975. As a result, the biological processes that controlled the way crops used the nutrients changed and the leached nitrogen from farmland soils has become a source of pollution.[8]

2.2.4 Organic farming

Organic farming is defined in different legal terms by different nations, but its main distinction from conventional farming is that it prohibits the use of synthetic chemicals in crop and livestock production. Often, it also includes diverse crop rotations and provides non-cropped habitat for insects that provide ecosystem services, such as pest control and pollination.[9] However, it is merely encouraged that organic farmers follow those kinds of wildlife friendly practices, and as a result there is a great difference between the ecosystem services that similarly sized but distinctly managed organic farms provide.[10] A recent review of the 76 studies concerning the relationship between biodiversity and organic farming listed three practices associated with organic farming that accounted for the higher biodiversity counts found in organic farms as compared to conventional farms.

"1. *Prohibition/reduced use of chemical pesticides and inorganic fertilizers* is likely to have a positive impact through the removal of both direct and indirect negative effects on arable plants, invertebrates and vertebrates.
2. *Sympathetic management of non-crop habitats and field margins* can enhance diversity and abundance of arable plants, invertebrates, birds and mammals.

3. *Preservation of mixed farming* is likely to positively impact farmland biodiversity through the provision of greater habitat heterogeneity at a variety of temporal and spatial scales within the landscape.[11]"

2.3 See also

- Agroecology
- Landscape of agriculture
- Regenerative agriculture

2.4 Notes

[1] 1.^ Jackson et al., The Farm as Natural Habitat, Introduction

[2] ^ Macdonald, Key Topics in Conservation Biology, Chapter 16

[3] 3.^ http://www.fao.org/newsroom/en/focus/2004/51102/index.html

[4] 4. Jackson et al., The Farm as Natural Habitat, Ch. 10

[5] 5.^ Benton et al., 182

[6] 6.^ Benton et al., 183–184

[7] 7.^ Macdonald et al., Key Topics in Conservation Biology, Ch 4

[8] 8. Jackson et al., The Farm as Natural Habitat, Ch. 10

[9] 9.^ Zhang et al., 255

[10] 10.^ Hole D.G. et al., 114

[11] 11.^ Hole D.G. et al., 120

2.5 References

- Altieri, Miguel A. 1999. The ecological role of biodiversity in agroecosystems: Agriculture, Ecosystemsand Environment 74: 19–31.

- Benton, Tim G., Vickery, Juliet A., Wilson, Jeremy D. 2003. Farmland biodiversity: is habitat heterogeneity the key? Trends in Ecology and Evolution 18: 182–188

- Dabbert, Stephan, 2002, Organic Agriculture and the Environment. OECD Publications Service

- FAO, http://www.fao.org/newsroom/en/focus/2004/51102/index.html

- Fiedler, Anna K., Landis, Douglas A., Wratten, Steve D. 2008. Maximizing ecosystem services from conservation biological control: The role of habitat management. Biological Control 45: 254–271

- Hole. D.G., Perkins, A.J., Wilson, D.J., Alexander, I.H., Grice, P.V., Evans, A.D. 2005. Biological Conservation 112:113–130

- Jackson, Dana L, Jackson, Laura L. 2002. The Farm as Natural Habitat. Island Press, Washington.

- Leopold, Aldo. 1939. The Farmer as a Conservationist. Pages 255–265 in Flader, Susan L., Callicott, J. Baird, editors. The River of the Mother of God. University of Wisconsin Press.

- Macdonald, David W., Service, Katrina. 2007. Key Topics in Conservation Biology. Blackwell Publishing, Oxford.

- Schmidt, Martin H. Tscharntke, Teja. 2005. The role of perennial habitats for Central European farmland spiders. Agriculture, Ecosystems and Environment 105: 235–242

- Shannon, D., Sen, A.M., Johnson, D.B. 2002. A comparative study of the microbiology of soils managed under organic and conventional regimes. Soil Use and Management 18: 274–283

- Zhang, Wei., Rickets, Taylor H., Kremen, Claire., Carney, Karen., Swinton, Scott M. 2007. Ecosystem services and dis-services to agriculture. Ecological Economics 64: 253–260

Chapter 3

Arborloo

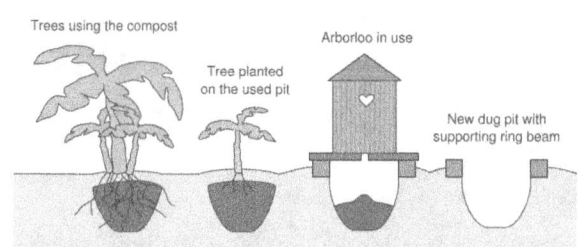

Steps of usage of the arborloo.

Arborloo in Ekwendeni, Malawi

An **Arborloo** is a simple type of dry toilet consisting of a pit (like a pit latrine, but less deep), concrete slab, superstructure (toilet house or outhouse) to provide privacy and possibly a ring beam to protect the pit from collapsing.

The concept of the arborloo toilet is to collect feces in a pit, and subsequently to grow a tree such as a fruiting tree in this very fertile soil.

The arborloo works by temporarily putting the slab and superstructure above a shallow pit while this pit fills. When the pit is nearly full, the superstructure and slab is moved to a newly dug pit and the old pit is covered with the earth got by digging the new pit and left to compost. The old site uses a bed for fruit tree or other, which is preferably planted during the rainy season.[1]

The arborloo can be considered a simple form of composting toilet. In using the nutrient-rich soil of a retired pit, the arborloo in effect treats feces as a resource rather than a waste product.

3.1 Design

The defecation pit may be circular or square and this may depend on the slab and superstructe. A circular pit is less likely to collapse.[2] The pit of the arborloo is shallow (between 1-1.5 meter).[1]

If the pit is dug by hand it must have a diameter of at least 0.9 meters to accommodate effective digging.[3] The pit should not be wider than the slab and must allow for 0.1 meter bearing around the edge.[3]

3.2 See also

- Ecological sanitation
- Reuse of excreta
- Treebog, another version

3.3 References

[1] Morgan, Peter (2007). *Toilets that make compost: Low-cost, sanitary toilets that produce valuable compost for crops in an African context.* Stockholm: EcoSanRes Programme. ISBN 978-9-197-60222-8.

[2] WEDC (2012). *An engineer's guide to latrine slabs* (PDF). Loughborough University: WEDC. p. 4. ISBN 978 1 84380 143 6.

[3] CAWST (2011). *Introduction to Low Cost Sanitation Latrine Construction.* CAWST: Center for Affordable Water and Sanitation Technology. p. 16.

3.4 External links

Media related to Arborloo at Wikimedia Commons

Chapter 4

Companion planting

Companion planting of carrots and onions

Companion planting in gardening and agriculture is the planting of different crops in proximity for pest control, pollination, providing habitat for beneficial creatures, maximizing use of space, and to otherwise increase crop productivity.[1] Companion planting is a form of polyculture.

Companion planting is used by farmers and gardeners in both industrialized and developing countries for many reasons. Many of the modern principles of companion planting were present many centuries ago in cottage gardens in England and forest gardens in Asia, and thousands of years ago in Mesoamerica.

4.1 History

In China, mosquito ferns (*Azolla* spp.) have been used for at least a thousand years as companion plants for rice crops. They host a cyanobacterium that fixes nitrogen from the atmosphere, and they block light from plants that would compete with the rice.[2]

Companion planting was practiced in various forms by the indigenous peoples of the Americas prior to the arrival of Europeans. These peoples domesticated squash 8,000 to 10,000 years ago,[3][4] then maize, then common beans, forming the Three Sisters agricultural technique. The cornstalk served as a trellis for the beans to climb, and the beans fixed nitrogen, benefitting the maize.[5][6][7]

Companion planting was widely promoted in the 1970s as part of the organic gardening movement. It was encouraged for pragmatic reasons, such as natural trellising, but mainly with the idea that different species of plant may thrive more when close together. It is also a technique frequently used in permaculture, together with mulching, polyculture, and changing of crops.

4.2 Examples of companion plants

See also: List of companion plants, List of beneficial weeds, and List of pest-repelling plants

Nasturtium (*Tropaeolum majus*) is a food plant of some caterpillars which feed primarily on members of the cabbage family (brassicas),[8] and some gardeners claim that planting them around brassicas protects the food crops from damage, as eggs of the pests are preferentially laid on the nasturtium. This practice is called trap cropping (using alternative plants to attract pests away from a main crop). However, while many trap crops have successfully diverted pests off of focal crops in small scale greenhouse, garden and field experiments, only a small portion of these plants have been shown to reduce pest damage at larger commercial scales.[9]

The smell of the foliage of marigolds is claimed to deter aphids from feeding on neighbouring crops. Marigolds with simple flowers also attract nectar-feeding adult hoverflies, the larvae of which are predators of aphids.

Various legume crops benefit from being commingled with a grassy nurse crop. For example, common vetch or hairy vetch is planted together with rye or winter wheat to make a good cover crop or green manure (or both).

The terms "undersowing" and "overseeding" both involve intercropping as a type of companion planting. "Undersowing" conveys the idea of sowing the second crop among the young plants of the first crop (or in between the rows, if rows are used). A connotation of understory growth is conveyed, albeit exaggerated (because the first crop is not yet a dense canopy). "Overseeding" conveys the idea of broadcasting the seeds of the second crop over the existing first crop. This is analogous to overseeding a lawn to improve the mix of grasses present.

4.3 Versions

There are a number of systems and ideas using companion planting.

Square foot gardening attempts to protect plants from many normal gardening problems by packing them as closely together as possible, which is facilitated by using companion plants, which can be closer together than normal.

Another system using companion planting is the forest garden, where companion plants are intermingled to create an actual ecosystem, emulating the interaction of up to seven levels of plants in a forest or woodland.

Organic gardening may make use of companion planting, since many synthetic means of fertilizing, weed reduction and pest control are forbidden.

4.3.1 Host-finding disruption

Recent studies on host-plant finding have shown that flying pests are far less successful if their host-plants are surrounded by any other plant or even "decoy-plants" made of green plastic, cardboard, or any other green material.

The host-plant finding process occurs in phases:

- The first phase is stimulation by odours characteristic to the host-plant. This induces the insect to try to land on the plant it seeks. But insects avoid landing on brown (bare) soil. So if only the host-plant is present, the insects will quasi-systematically find it by simply landing on the only green thing around. This is called (from the point of view of the insect) "appropriate landing". When it does an "inappropriate landing", it flies off to any other nearby patch of green. It eventually leaves the area if there are too many 'inappropriate' landings.

- The second phase of host-plant finding is for the insect to make short flights from leaf to leaf to assess the plant's overall suitability. The number of leaf-to-leaf

flights varies according to the insect species and to the host-plant stimulus received from each leaf. The insect must accumulate sufficient stimuli from the host-plant to lay eggs; so it must make a certain number of consecutive 'appropriate' landings. Hence if it makes an 'inappropriate landing', the assessment of that plant is negative, and the insect must start the process anew.

Thus it was shown that clover used as a ground cover had the same disruptive effect on eight pest species from four different insect orders. An experiment showed that 36% of cabbage root flies laid eggs beside cabbages growing in bare soil (which resulted in no crop), compared to only 7% beside cabbages growing in clover (which allowed a good crop). Simple decoys made of green cardboard also disrupted appropriate landings just as well as did the live ground cover.[10]

4.4 Companion plant categories

The use of companion planting can be of benefit to the grower in a number of different ways, including:

- **Hedged investment** – the growing of different crops in the same space increases the odds of some yield being given, even if one crop fails.

- **Increased level interaction** – when crops are grown on different levels in the same space, such as providing ground cover or one crop working as a trellis for another, the overall yield of a plot may be increased.

- **Protective shelter** is when one type of plant may serve as a wind break or provide shade for another.

- **Pest suppression** – some companion plants may help prevent pest insects or pathogenic fungi from damaging the crop, through chemical means.[11]

- **Predator recruitment** and **positive hosting** – The use of companion plants that produce copious nectar or pollen in a vegetable garden (insectary plants) may help encourage higher populations of beneficial insects that control pests,[12] as some beneficial predatory insects only consume pests in their larval form and are nectar or pollen feeders in their adult form.

- **Trap cropping** – some companion plants are claimed to attract pests away from others.

- **Pattern disruption** – in a monoculture pests spread easily from one crop plant to the next, whereas such easy progress may be disrupted by surrounding companion plants of a different type.[10]

4.5   See also

- Satoyama

- Intercropping

- Ecological facilitation

- Vegan organic gardening

4.6   References

[1] McClure, Susan (1994). *Companion Planting*. Rodale Press. ISBN 0-87596-616-0.

[2] "Plant Resources for Human Development-Nitrogen in Rice" (PDF). Dhakai.com. Retrieved February 21, 2015.

[3] Smith, B. D. (1997). The initial domestication of *Cucurbita pepo* in the Americas 10,000 years ago. *Science* 276 932-34.

[4] "Cucurbitaceae--Fruits for Peons, Pilgrims, and Pharaohs". University of California at Los Angeles. Retrieved September 2, 2013.

[5] Mt. Pleasant, J. (2006). "The science behind the Three Sisters mound system: An agronomic assessment of an indigenous agricultural system in the northeast". In Staller, J. E.; et al. *Histories of maize: Multidisciplinary approaches to the prehistory, linguistics, biogeography, domestication, and evolution of maize*. Amsterdam. pp. 529–537.

[6] Landon, Amanda J. (2008). "The "How" of the Three Sisters: The Origins of Agriculture in Mesoamerica and the Human Niche". *Nebraska Anthropologist*. Lincoln, NE: University of Nebraska-Lincoln: 110–124.

[7] Bushnell, G. H. S. (1976). "The Beginning and Growth of Agriculture in Mexico". *Philosophical Transactions of the Royal Society of London*. London: Royal Society of London. **275** (936): 117–120. doi:10.1098/rstb.1976.0074.

[8] "Cabbage caterpillars". Royal Horticultural Society. Retrieved 10 February 2013.

[9] Holden, Matthew H.; Ellner, Stephen P.; Lee, Doo-Hyung; Nyrop, Jan P.; Sanderson, John P. (2012-06-01). "Designing an effective trap cropping strategy: the effects of attraction, retention and plant spatial distribution". *Journal of Applied Ecology*. **49** (3): 715–722. doi:10.1111/j.1365-2664.2012.02137.x. ISSN 1365-2664.

[10] Horticulture Research International, Wellesbourne : "Insects can see clearly now the weeds have gone". Finch, S. & Collier, R. H. (2003). Biologist, 50 (3), 132-135

[11] "The Self-Sufficient Gardener Podcast--Episode 24 Companion Planting and Crop Rotation". Retrieved 2010-08-13.

[12] "Pacific Northwest Nursery IPM. Flowers, Sweets and a Nice Place to Stay: Courting Beneficials to Your Nursery". Oregon State University. Retrieved 11 February 2013.

4.7   External links

- National Gardening Association - *Companion and Interplanting*, with companion planting table

Chapter 5

Composting toilet

Composting toilet at Activism Festival 2010 in the mountains outside Jerusalem

A **composting toilet** is a type of dry toilet that uses a predominantly aerobic processing system to treat human excreta, by composting or managed aerobic decomposition. These toilets generally use little to no water and may be used as an alternative to flush toilets.[1] They have found use in situations where no suitable water supply or sewer system and sewage treatment plant is available to capture the nutrients in human excreta. They are in use in many roadside facilities and national parks in Sweden, Canada, US, UK and Australia. They are used in rural holiday homes in Sweden and Finland.

The human excreta is usually mixed with sawdust, coconut coir or peat moss to facilitate aerobic processing, liquid absorption, and odor mitigation. Most composting toilets use slow, cold composting conditions, sometimes connected to a secondary external composting step.

Composting toilets produce a compost that may be used for horticultural or agricultural soil enrichment if the local regulations allow this. A curing stage is often needed to allow mesophilic composting to reduce potential phytotoxins.

5.1 Terminology

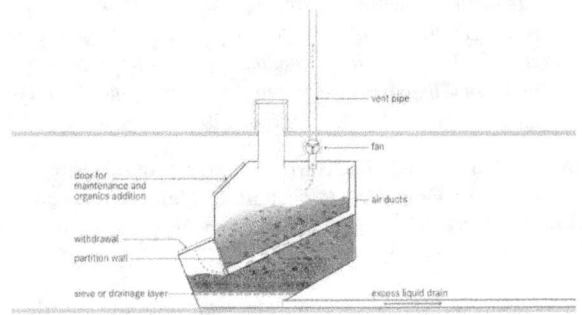

Schematic of the composting chamber which is located below the toilet seat[1]

The term "composting toilet" is used quite loosely, and its meaning may vary by country. For example, in English-speaking countries, the term "anaerobic composting" (equivalent to anaerobic decomposition) is used. In Germany and Scandinavian countries, composting always refers to a predominantly aerobic process. This aerobic composting may take place with an increase in temperature due to microbial action, or without a temperature increase in the case of slow composting or cold composting. If earth worms are used (vermicomposting) then there is also no increase in temperature.

Composting toilets differ from pit latrines, arborloo or tree bogs, all of which are forms of less controlled decomposition and may not protect groundwater from nutrient or pathogen contamination or provide optimal nutrient recycling. They also differ from urine-diverting dry toilets (UD-

DTs) where pathogen reduction is achieved through dehydration (also known by the more precise term "desiccation") and where the faeces collection vault is kept as dry as possible. Composting toilets target a certain degree of moisture in the composting chamber.

Composting toilets usually do not divert urine. Offering a waterless urinal in addition to the toilet can help keep excess amounts of urine out of the composting chamber.

Composting toilets can be used to implement an ecological sanitation approach for resource recovery, and some people call their composting toilet designs "ecosan toilets" for that reason. However, this is not recommended as the two terms (i.e. composting and ecosan) are not identical.[2][3]

Composting toilets have also been called "sawdust toilets", which can be appropriate if the amount of aerobic composting taking place in the toilet's container is very limited.[4] The "Clivus multrum" is a type of composting toilet which has a large composting chamber below the toilet seat and also receives undigested organic material to increase the carbon to nitrogen ratio.

5.2   Applications

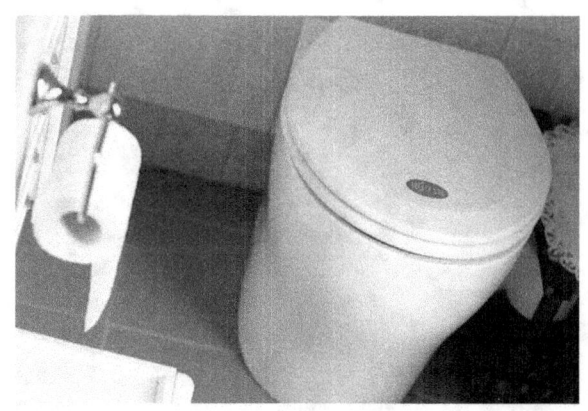

This is the pedestal for a split-system composting toilet where collection/treatment chambers are located below the bathroom floor.

Composting toilets can be suitable in areas such as a rural area or a park that lacks a suitable water supply, sewers and sewage treatment. They can also help increase the resilience of existing sanitation systems in the face of possible natural disasters such as climate change, earthquakes or tsunami. Composting toilets can reduce or perhaps eliminate the need for a septic tank system to reduce environmental footprint (particularly when used in conjunction with an on-site greywater treatment system).

These types of toilets can be used for resource recovery by reusing sanitized feces and urine as fertilizer and soil conditioner for gardening or ornamental activities.

Inexpensive do-it-yourself compost toilet at Dial House, Essex, England, utilizing an old desk as the toilet unit.

Public composting toilet at a highway rest facility in Sweden

5.3   Basics

Main article: Compost

5.3.1   Components

A composting toilet consists of two elements: a place to sit or squat and a collection/composting unit.[2] The composting unit consists of four main parts:[1]

- storage or composting chamber

- a ventilation unit to ensure that the degradation process in the toilet is predominantly aerobic and to vent odorous gases

- a leachate collection system to remove excess liquid

- an access door for extracting the compost

5.3.2 Construction

The composting chamber can be constructed above or below ground level. It can be inside a structure or include a separate superstructure.

A drainage system removes leachate. Otherwise, excess moisture can cause anaerobic conditions and impede decomposition. Urine diversion can improve compost quality, since urine contains large amounts of ammonia that inhibits microbiological activity.[5]

Composting toilets greatly reduce human waste volumes through psychrophilic, thermophilic or mesophilic composting. Keeping the composting chamber insulated and warm protects the composting process from slowing due to low temperatures.

5.3.3 Odorous gases

The following gases may be emitted during the composting process that takes place in composting toilets: hydrogen sulfide (H_2S), ammonia, nitrous oxide (N_2O) and volatile organic compounds (VOCs).[6] These gases can potentially lead to complaints about odours. Some methane may also be present, but it is not odorous.

5.4 Pathogen removal

Excreta-derived compost recycles fecal nutrients, but it can carry and spread pathogens if the process of reuse of excreta is not done properly.

Internal pathogen destruction rates are usually low, particularly helminth eggs, such as Ascaris eggs.[4] This carries the risk of spreading disease if a proper system management is not in place. Compost from human excreta processed under only mesophilic conditions or taken directly from the compost chamber is not safe for food production.[7] High temperatures or long composting times are required to kill helminth eggs, the hardiest of all pathogens. Helminth infections are common in many developing countries.

In thermophilic composting bacteria that thrive at temperatures of 40–60 °C (104–140 °F) oxidize (break down) waste into its components, some of which are consumed in the process, reducing volume and eliminating potential pathogens. To destroy pathogens, thermophilic composting must heat the compost pile sufficiently, or enough time (1–2 years) must elapse since fresh material was added that biological activity has had the same pathogen removal effect.

One guideline claims that pathogen levels are reduced to a safe level by thermophilic composting at temperatures of 55 °C for at least two weeks or at 60 °C for one week.[2] An alternative guideline claims that complete pathogen destruction may be achieved already if the entire compost heap reaches a temperature of 62 °C (144 °F) for one hour, 50 °C (122 °F) for one day, 46 °C (115 °F) for one week or 43 °C (109 °F) for one month,[5] although others regard this as overly optimistic.[2]

5.5 Design considerations

Composting toilet with a seal in the lid in Germany

5.5.1 Environmental factors

Four main factors affect the decomposition process:[5]

- Sufficient oxygen is necessary for aerobic composting

- Moisture content from 45 to 70 percent (heuristically, "the compost should feel damp to the touch, with only a drop or two of water expelled when tightly squeezed in the hand."[2])

- Temperature between 40 and 50 °C (achieved through proper chamber dimensioning and possibly active mixing)

- Carbon-to-nitrogen ratio (C:N) of 25:1

5.5.2 Additives and bulking material

Human excreta and food waste do not provide optimum conditions for composting. Usually the water and nitrogen content is too high, particularly when urine is mixed with feces. Additives or "bulking material", such as wood chips, bark chips, sawdust, ash and pieces of paper can absorb moisture. The additives improve pile aeration and increase the carbon to nitrogen ratio.[2] Bulking material also covers faeces and reduces insect access. Absent sufficient bulking material, the material may become too compact and form impermeable layers, which leads to anaerobic conditions and odour.[2]

5.5.3 Leachate management

Leachate removal controls moisture levels, which is necessary to ensure rapid, aerobic composting. Some commercial units include a urine-separator or urine-diverting system and/or a drain at the bottom of the composter for this purpose.

5.5.4 Aeration and mixing

Microbial action also requires oxygen, typically from the air. Commercial systems provide ventilation that moves air from the bathroom, through the waste container, and out a vertical pipe, venting above the roof. This air movement (via convection or fan forced) passes carbon dioxide and odors.

Some units require manual methods for periodic aeration of the solid mass such as rotating the composting chamber or pulling an "aerator rake" through the mass.

5.6 Types

Commercial units and construct-it-yourself systems are available.[8] Variations include number of composting vaults, removable vault, urine diversion and active mixing/aeration.[2]

External composting chamber of a composting toilet at a house in France

5.6.1 Slow composting (or moldering) toilets

Most composting toilets use slow composting which is also called "cold composting". The compost heap is built up step by step over time.

The finished end product from "slow" composting toilets ("moldering toilets" or "moldering privies" in the US), is generally not free of pathogens. World Health Organization Guidelines from 2006 offer a framework for safe reuse of excreta, using a multiple barrier approach.[9]

Slow composting toilets employ a passive approach. Common applications involve modest and often seasonal use, such as remote trail networks. They are typically designed such that the materials deposited can be isolated from the operational part. The toilet can also be closed to allow further mesophilic composting.[10] Slow composting toilets rely on long retention times for pathogen reduction and for decomposition of excreta or on the combination of time and/or the addition of red wriggler worms for vermicomposting. Worms can be introduced to accelerate composting. Some jurisdictions of the US consider these worms as invasive species.[9]

Example in Vermont woods

Slow composting toilets have been installed by the Green Mountain Club in Vermont's woodlands. They employ multiple vaults (called cribs) and a movable building. When one of the vaults fills, the building is moved over an empty vault. The full vault is left untouched for as long as possible (up to three years) before it is emptied. The large surface area and exposure to air currents can cause the pile to dry out. To counteract this, signs instruct users to urinate in the toilet.[11] The club also uses pit latrines and simple bucket toilets with woodchips and external composting and directs users to urinate in the forest to prevent odiferous anaerobic conditions.[12]

5.6.2 Active composters

Self-contained

"Self-contained" composting toilets compost in a container within the toilet unit. They are slightly larger than a flush toilet, but use roughly the same floor space. Some units use fans for aeration, and optionally, heating elements to maintain optimum temperatures to hasten the composting process and to evaporate urine and other moisture. Operators of composting toilets commonly add a small amount of absorbent carbon material (such as untreated sawdust, coconut coir, peat moss) after each use to create air pockets to encourage aerobic processing, to absorb liquid and to create an odor barrier. This additive is sometimes referred to as "bulking agent." Some owner-operators use microbial "starter" cultures to ensure composting bacteria are in the process, although this is not critical.

Remote

"Remote" "central" or "underfloor" units collect excreta via a toilet stool, either waterless, vacuum or micro-flush, from which it drains into a composter. "Vacuum-flush systems" can flush horizontally or upward with a small amount of water to the composter. "Micro-flush" toilets use about 500 millilitres (17 US fl oz) per use. These units feature a chamber below the toilet stool (such as in a basement or outside) where composting takes place and are suitable for high-volume and year-round applications as well as to serve multiple toilet stools.[13]

5.6.3 Other

Some units employ roll-away containers fitted with aerators, while others use sloped-bottom tanks.

5.7 Maintenance

Maintenance is critical to ensure proper operation, including odor prevention. Maintenance tasks include: cleaning, servicing technical components such as fans and removal of compost, leachate and urine. Urine removal is only required for those types of composting toilets using urine diversion.

Once composting is complete (or more often), the compost must be removed from the unit. How often this occurs is a function of container size, usage and composting conditions, such as temperature.[2] Active, hot composting may span months only while passive, cold composting may require years. Properly managed units yield output volumes of about 10% of inputs.

5.8 Uses of compost

Finished compost from a composting toilet ready for application as soil improvement in Kiel-Hassee, Germany

Main articles: Uses of compost and Reuse of excreta

The material from composting toilets is a humus-like material, which can be suitable as a soil amendment for agriculture. Compost from residential composting toilets can be used in domestic gardens, and this is the main such use.

Enriching soil with compost adds substantial nitrogen, phosphorus, potassium, carbon and calcium. In this regard compost is equivalent to many fertilizers and manures purchased in garden stores. Compost from composting toilets has a higher nutrient availability than the dried faeces that result from a urine-diverting dry toilet.[2]

Urine is typically present, although some is lost via leaching and evaporation. Urine can contain up to 90 percent of the residual nitrogen, up to 50 percent of the phosphorus, and up to 70 percent of the potassium.[14]

Compost derived from these toilets has in principle the same uses as compost derived from other organic waste products, such as sewage sludge or municipal organic waste. However, users of excreta-derived compost must consider the risk of pathogens.

5.8.1 Pharmaceutical residues

Excreta-derived compost may contain prescription pharmaceuticals. Such residues are also present in conventional wastewater treatment effluent. This could contaminate groundwater. Among the medications that have been found in groundwater in recent years are antibiotics, antidepressants, blood thinners, ACE inhibitors, calcium-channel blockers, digoxin, estrogen, progesterone, testosterone, Ibuprofen, caffeine, carbamazepine, fibrates and cholesterol-reducing medications.[15] Between 30% and 95% of pharmaceuticals medications are excreted by the human body. Medications that are lipophilic (dissolved in fats) are more likely to reach groundwater by leaching from fecal wastes. Wastewater treatment plants remove an average of 60% of these medications.[16] The percentage of medications degraded during composting of excreta has not yet been reported.

5.9 Comparison

5.9.1 Pit latrines

Main article: Pit latrine

Unlike pit latrines, composting toilets convert feces into a dry, odorless material, avoiding the issues surrounding liquid fecal sludge management (e.g. odor, insects and disposal). These toilets minimize the risk of water pollution through the safe containment of feces in above-ground vaults, which allows the toilets to be sited in locations where pit-based systems are not appropriate.

However, composting toilets face higher capital costs (although lifecycle costs might be lower) and greater complexity (for instance, adding covering materials and managing moisture content).

5.9.2 Flush toilets

Main article: Flush toilet

Unlike flush toilets, composting toilets do not dilute excreta and create wastewater streams which must be treated before disposal. On the other hand, wastewater treatment plants can centralize waste management for an entire community, with potentially greater efficiency.

5.9.3 Urine-diverting dry toilets

Main article: Urine-diverting dry toilet

Composting toilets are more difficult to maintain than other types of dry toilets, like urine-diverting dry toilets (UDDT) with which they are often confused. This is due to the need to maintain a consistent and relatively high moisture content, as well as the relatively high complexity of composting toilets compared to UDDTs. Apart from that, composting toilets are quite similar to UDDTs, sharing many of the same advantages and disadvantages.

5.10 History

5.10.1 Dry earth toilet

Before the flush toilet became accepted in the late 19th century in developed countries, some inventors, scientists and public health officials supported the use of "dry earth closets", a type of dry toilet with similarities to composting toilets, but the collection vessel for the human excreta was not designed to compost. Dry earth closets were invented by English clergyman Henry Moule, who dedicated his life to improving public sanitation after witnessing the cholera epidemics of 1849 and 1854. Impressed by the insalubrity of the houses, especially during the Great Stink in the summer of 1858, he invented what he called the 'dry earth system'.

In partnership with James Bannehr, he patented his device (No. 1316, dated 28 May 1860). Among his works bearing on the subject were *The Advantages of the Dry Earth System* (1868), *The Impossibility overcome: or the Inoffensive, Safe, and Economical Disposal of the Refuse of Towns and Villages* (1870), The *Dry Earth System* (1871), *Town Refuse, the Remedy for Local Taxation* (1872), and *National Health and Wealth promoted by the general adoption of the Dry Earth System* (1873).

His system was adopted in private houses, in rural districts, in military camps, in many hospitals, and extensively in the British Raj. Ultimately, however, it failed to gain public

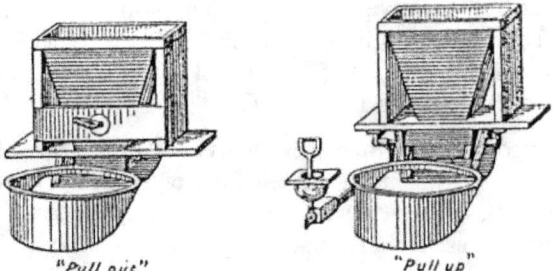

Figs. 8 and 9.—Moule's Earth Closets.

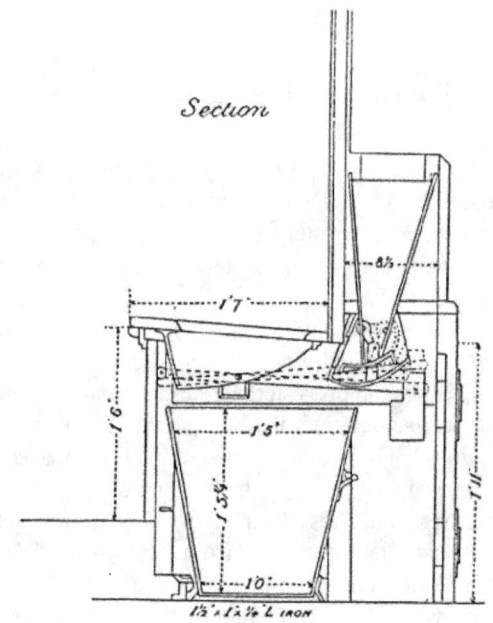

Fig. 10.—Section of Moule's Automatic Earth Closet.

Henry Moule's earth closet, patented in 1873 (not a true composting toilet). Example from around 1875. Rear chamber for dispensing cover material

Henry Moule's earth closet design, circa 1909.

support as attention turned to the water-flushed toilet connected to a sewer system.

In Germany, a similar dry toilet with a peat dispenser was marketed until after the second World War (it was called "Metroclo" and was manufactured by Gefinal, Berlin).

5.11 Society and culture

5.11.1 Regulations

International Organization for Standardization (ISO)

The International Organization for Standardization (ISO) is currently preparing a "management standard". As of 2015 this was in a draft state as ISO 24521, under the heading "Activities relating to drinking water and wastewater services — Guidelines for the management of basic onsite domestic wastewater services".[17] The standard is meant to be used in conjunction with ISO 24511.[18] It deals with toilets

(including composting toilets) and toilet waste. The guidelines are applicable to basic wastewater systems and include the complete domestic wastewater cycle, such as planning, usability, operation and maintenance, disposal, reuse and health.

International Association of Plumbing and Mechanical Officials

The International Association of Plumbing and Mechanical Officials (IAPMO) is a plumbing and mechanical code structure adopted by many developed countries. It recently proposed an addition to its "Green Plumbing Mechanical Code Supplement" that, "...outlines performance criteria for site built composting toilets with and without urine diversion and manufactured composting toilets."[19] If adopted, this composting and urine diversion toilet code (the first of its kind in the United States) will appear in the 2015 edition of the Green Supplement to the Uniform

Plumbing Code.[20][21]

United States

No performance standards for composting toilets are universally accepted in the US. Seven jurisdictions in North America[22] use *American National Standard/NSF International Standard ANSI/NSF 41-1998: Non-Liquid Saturated Treatment Systems*. An updated version was published in 2011.[23][note 1] Systems might also be listed with the Canadian Standards Association, cETL-US, and other standards programs.

Regarding byproduct regulation, several US states permit disposal of solids from composting toilets (usually a distinction between different types of dry toilets is not made) by burial, with varying or no minimum depth mandates (as little as 6 inches). For instance:

- Massachusetts: "Residuals from the composting toilet system must be buried on-site and covered with a minimum of six inches of clean compacted soil.[24] Massachusetts requires that any liquids produced but, "not recycled through the toilet [itself be] either discharged through a greywater system on the property that includes a septic tank and soil absorption system, or removed by a licensed septage hauler."[24]

- Oregon: "Humus from composting toilets may be used around ornamental shrubs, flowers, trees, or fruit trees and shall be buried under at least twelve inches of soil cover."[25]

- Rhode Island: "Solids produced by alternative toilets may be buried on site," while, "residuals shall not be applied to food crops."[26]

- Virginia: "All materials removed from a composting privy shall be buried," and "compost material shall not be placed in vegetable gardens or on the ground surface."[27]

- Vermont: "Byproducts may be disposed via "...shallow burial in a location approved by the Agency that meets the minimum site conditions [required for an onsite septic tank-based sanitation system]."[28]

- Washington: models its extensive regulations for what it refers to as "waterless toilets" on the federal regulations that govern sewage sludge.[29]

The Environmental Protection Agency has no jurisdiction over the byproducts of a dry toilet as long as excreta are not referred to as "fertilizer" (but instead simply a material that is being disposed of). Federal rule 503, known colloquially as the "EPA Biosolids rule" or the "EPA sludge rule" applies only to fertilizer. Thus, individual states regulate composting toilets.[30][31]

Germany

The regulations for composting toilets and other forms of dry toilets in Germany vary from state to state and from one application to another (e.g. use in allotment gardens or use in family homes and settlements). In the different states of Germany, it is the "Landesbauordnung" (translates to "state civil engineering regulations") of the respective state that regulates the use of such alternative toilets.[32] Most of them stipulate the use of flush toilets, however there are many exceptions, for example in the states of Hamburg, Lower Saxony, Bavaria, Mecklenburg-Western Pomerania, Rhineland-Palatinate, Saxony-Anhalt and Thuringia.[32] These generally make exceptions for the use of composting toilets in homes provided that there are no concerns for public health.

Regulations governing the use of compost and urine from composting toilets is less clear in Germany but it seems generally allowed provided it is used on one's own property and not sold to third parties.[32]

5.11.2 Examples

Finland

Numerous sparsely settled villages in rural areas in Finland are not connected to municipal water supply or sewer networks, requiring homeowners to operate their own systems. Individual private wells, i.e. shallow dug wells or boreholes in the bedrock, are often used for water supply, and many homeowners have opted for composting toilets. In addition, these toilets are common at holiday homes, often located near sensitive water bodies. For these reasons, many manufacturers of composting toilets are based in Finland, including Biolan, Ekolet, Kekkilä, Pikkuvihreä and Raita Environment.[33][34]

Estimates made by leading Finnish composting toilet manufacturers and the Global Dry Toilet Association of Finland provided the following 2014 figures for composting toilet use in Finland:

- About 4% of single-family homes not connected to a public sewer network are equipped with a composting toilet.

- Some 200,000 manufactured composting toilets are thought to serve holiday homes, matched by the number of other dry toilets. The simplest ones are sited in an outhouse.

Germany

Composting container of "TerraNova" composting toilet, showing open removal chamber (town house at the ecological settlement Hamburg-Allermöhe, Germany)

Composting toilets have been successfully installed in houses with up to four floors.[2] An estimate from 2008 put the number of composting toilets in households in Germany at 500.[35] Most of these residences are also connected to a sewer system; the composting toilet was not installed due a lack of sewer system but for other reasons, mainly because of an "ecological mindset" of the owners.

In Germany and Austria, composting toilets and other types of dry toilets have been installed in single and multi-family houses (e.g. Hamburg, Freiburg, Berlin), ecological settlements (e.g. Hamburg-Allemöhe, Hamburg-Braamwisch, Kiel-Hassee, Bielefeld-Waldquelle, Wien-Gänserndorf) and in public buildings (e.g. Ökohaus Rostock, VHS-Ökostation Stuttgart-Wartberg, public toilets in recreational areas, restaurants and huts in the Alps, house boats and forest Kindergartens).[35]

The ecological settlement in Hamburg-Allermöhe has had composting toilets since 1982. The settlement of 36 single-family houses with approximately 140 inhabitants uses composting toilets, rainwater harvesting and constructed wetlands. Composting toilets save about 40 litres of water per capita per day compared to a conventional flush toilet (10 liter per flush), which adds up to 2,044 m^3 water savings per year for the whole settlement.[36]

Worldwide

Composting toilets with a large composting container (of the type Clivus Multrum and derivations of it) are popular in US, Canada, Australia, New Zealand and Sweden. They can be bought and installed as commercial products, as designs for self builders or as "design derivatives" which are marketed under various names. It has been estimated that approximately 10,000 such toilets might be in use worldwide.

5.12 See also

- Bucket toilet

5.13 Notes

[1] A listing of the most current NSF/ANSI standards can be found in PDF format at NSF International's *Standards* subdomain.

5.14 References

[1] Tilley, E.; Ulrich, L.; Lüthi, C.; Reymond, Ph.; Zurbrügg, C. *Compendium of Sanitation Systems and Technologies - (2nd Revised Edition)*. Swiss Federal Institute of Aquatic Science and Technology (Eawag), Duebendorf, Switzerland. p. 72. ISBN 978-3-906484-57-0.

[2] Berger, W. (2011). Technology review of composting toilets - Basic overview of composting toilets (with or without urine diversion). Deutsche Gesellschaft für Internationale Zusammenarbeit (GIZ) GmbH, Eschborn, Germany

[3] Rieck, C., von Münch, E., Hoffmann, H. (2012). Technology review of urine-diverting dry toilets (UDDTs) - Overview on design, management, maintenance and costs. Deutsche Gesellschaft fuer Internationale Zusammenarbeit (GIZ) GmbH, Eschborn, Germany

[4] Hill, B. G. (2013). An evaluation of waterless human waste management systems at North American public remote sites. PhD thesis, University of British Columbia (Vancouver), Canada

[5] "The online Compendium of Sanitation Systems and Technologies". *The online Compendium of Sanitation Systems*

and Technologies. eawag aquatic research. 2014. Retrieved 2014-12-29.

[6] Font, Xavier; Artola, Adriana; Sánchez, Antoni (6 April 2011). "Detection, Composition and Treatment of Volatile Organic Compounds from Waste Treatment Plants". *Sensors.* **11** (12): 4043–4059. doi:10.3390/s110404043.

[7] Stenström, T.A., Seidu, R., Ekane, N., Zurbrügg, C. (2011). Microbial exposure and health assessments in sanitation technologies and systems - EcoSanRes Series, 2011-1. Stockholm Environment Institute (SEI), Stockholm, Sweden, page 88

[8] National Small Flows Clearinghouse, West Virginia University, Composting toilet technology

[9] WHO (2006). WHO Guidelines for the Safe Use of Wastewater, Excreta and Greywater - Volume IV: Excreta and greywater use in agriculture. World Health Organization (WHO), Geneva, Switzerland

[10] Appalachian Trail Conservancy (2014). Backcountry Sanitation Manual, 2nd Edition. Appalachian Trail Conservancy, Green Mountain Club, USDA Forest Service, National Park Service, USA

[11] Allen, Lee (2013). "Long Trail News: Quarterly of the Green Mountain Club, Fall 2013. Article titled: "A Privy is a Privy is a Privy...or is it? To Pee or Not Pee."" (PDF). *Green Mountain Club.* Green Mountain Club. Retrieved 31 January 2013.

[12] Antos-Ketcham, Pete (2013). "Long Trail News: Quarterly of the Green Mountain Club, Fall 2013. Article titled: "Batch-Bin/Beyond-the-Bin (BTB) Composting Privies"" (PDF). *Green Mountain Club.* Green Mountain Club. Retrieved 31 January 2015.

[13] Berger, W. (2009). Appendix of technology review of composting toilets - List of manufacturers and commercially available composting toilets. Gesellschaft für Internationale Zusammenarbeit (GIZ) GmbH

[14] J.O. Drangert, Urine separation systems

[15] *Drugs in the Water.* Harvard Health Letter. 2011.

[16] Encyclopedia of Quantitative Risk Analysis and Assessment, Volume 1, edited by Edward L. Melnick, Brian S. Veritt, 2008

[17] "ISO/DIS 24521. Activities relating to drinking water and wastewater services -- Guidelines for the management of basic onsite domestic wastewater services". *International Organization for Standardization (ISO).* Retrieved 15 January 2015.

[18] "ISO 24511:2007. Activities relating to drinking water and wastewater services -- Guidelines for the management of wastewater utilities and for the assessment of wastewater services". *International Organization for Standardization (ISO).* Retrieved 15 January 2015.

[19] "Recode September 2014 Newsletter". *Recode.* Recode. September 2014. Retrieved 15 January 2015.

[20] "IAPMO Proposed Composting and Urine DIversion Toilet Code" (PDF). *The IAPMO Group.* International Association of Plumbing and Mechanical Officials. Retrieved 15 January 2015.

[21] Cole, Daniel (January 2015). "IAPMO GPMCS raising the bar for water, energy efficiency". *Plumbing Engineer.* Plumbing Engineer. Retrieved 15 January 2015.

[22] Oregon Onsite Advisory Committee "Final Report of Recommended Changes to Rules Governing Onsite Systems", *OR DEQ,* February 8, 2010, accessed May 8, 2011.

[23] "PUBLICATIONS - Standards and Criteria - March 21, 2013" (PDF). NSF International. p. 4. Retrieved 24 March 2013. Wastewater Treatment Units ... NSF/ANSI 41 – 2011: Non-liquid saturated treatment systems (composting toilets)

[24] "Regulatory Provisions for Composting Toilets and Greywater Systems". *The Official Website of the Massachusetts Executive Office of Energy and Environmental Affairs.* Office of Energy and Environmental Affairs. Retrieved 13 January 2015.

[25] "Department of Consumer and Business Services, Building Codes Division, Division 770, Plumbing Product Approvals". *Oregon Secretary of State.* State of Oregon. Retrieved 13 January 2015.

[26] "State of Rhode Island and Providence Plantations Department of Environmental Management, Office of Water Resources: "Rules Establishing Minimum Standards Relating to Location, Design, Construction and Maintenance of Onsite Wastewater Treatment Systems"" (PDF). *State of Rhode Island Department of Environmental Management.* STATE OF RHODE ISLAND AND PROVIDENCE PLANTATIONS. July 2010. Retrieved 13 January 2015.

[27] "SEWAGE HANDLING AND DISPOSAL REGULATIONS (Emergency Regulations for Gravelless Material and Drip Dispersal), 12 VAC 5-610-10 et seq." (PDF). *State of Virginia Department of Health.* Commonwealth of Virginia. 14 March 2014. Retrieved 13 January 2015.

[28] "Environmental Protection Rules, Chapter 1: Wastewater System and Potable Water Supply Rules" (PDF). *State of Vermont Drinking Water and Groundwater Protection Division.* State of Vermont. 29 September 2007. Retrieved 14 January 2015.

[29] "Recommended Standards and Guidance for Performance, Application, Design, and Operation & Maintenance: Water Conserving On-Site Wastewater Treatment Systems" (PDF). *State of Washington Department of Health.* State of Washington. July 2012. Retrieved 14 January 2015.

[30] "Water Efficiency Technology Fact Sheet: Composting Toilets" (PDF). *United States Environmental Protection Agency,*

Office of Water, Washington, D.C., EPA 832-F-99-066. United States Environmental Protection Agency, Office of Water. September 1999. Retrieved 13 January 2015.

[31] "TITLE 40—Protection of Environment, Chapter I—Environmental Protection Agency (Continued), Subchapter O—Sewage Sludge, Part 503—Standards for the Use or Disposal of Sewage Sludge". *Electronic Code of Federal Regulations.* United States Government Publishing Office. Retrieved 13 January 2015.

[32] Lorenz-Ladener, Hrsg. Claudia; Berger, Wolfgang (2005). *Kompost-Toiletten: Wege zur sinnvollen Fäkalienentsorgung* (1. überarb. u. erw. Aufl. ed.). Staufen im Breisgau: Ökobuch. p. 178. ISBN 978-3-936896-16-9.

[33] Global Dry Toilet Association of Finland (2011) Dry Toilet Manufacturers in Finland, Leaflet in English and Finnish

[34] "Global Dry Toilet Association of Finland". *Global Dry Toilet Association of Finland - Company and association members.* Retrieved 15 January 2015.

[35] Lorenz-Ladener, Hrsg. Claudia; Berger, Wolfgang (2005). *Kompost-Toiletten: Wege zur sinnvollen Fäkalienentsorgung* (1. überarb. u. erw. Aufl. ed.). Staufen im Breisgau: Ökobuch. p. 183. ISBN 978-3-936896-16-9.

[36] Rauschning, G., Berger, W., Ebeling, B., Schöpe, A. (2009). Ecological settlement in Allermöhe Hamburg, Germany - Case study of sustainable sanitation projects. Sustainable Sanitation Alliance (SuSanA)

5.15   External links

- "Compost Toilet Systems | NaturalToilets.com"

- "What is a Composting Toilet System and How Does it Compost?"

- Composting toilet description (Sustainable Sanitation and Water Management Toolbox)

- Composting systems (documents in library of Sustainable Sanitation Alliance)

- More photos of composting toilets in Flickr photo database of Sustainable Sanitation Alliance

Chapter 6

Ecosynthesis

see *Planetary ecosynthesis for Terraforming*

Ecosynthesis is the use of introduced species to fill niches in a disrupted environment, with the aim of increasing the speed of ecological restoration. This decreases the amount of physical damage done in a disrupted landscape.

6.1   See also

- Ecopoiesis

6.2   References

- Tane, Hakai (1995) *Ecography. Mapping and Modelling Landscape Ecosystems.* Canberra: The Murray-Darling Basin Commission.

- Holmgren, David (2002) *Permaculture. Principles and Pathways beyond Sustainability.* Hepburn, Victoria: Holmgren Design Services.

Chapter 7

Ernie and Erica Wisner

Ernie and Erica Wisner are a couple from Tonasket, WA best known for their innovative rocket mass heater designs. They are often referred to as the world wide leaders and trainers in rocket stove technology. They have made over 700 rocket stoves all over the world.[1]

7.1 Background

They were introduced to this type of work when Ernie apprenticed with Ianto Evans for two years. Ianto Evans, of Cob Cottage Co., was the designer who originally developed this type of technology 30 years ago. Erica and Ernie initially planned to share information about all different types of natural building but their fans demanded otherwise. Everyone wanted to be taught how to build the rocket mass heaters and the custom cob rocket stoves. So that is what they did.[2]

7.2 Current Affairs

Ernie and Erica now spend their time touring the world teaching workshops on how to build these energy efficient, natural heaters.[3] They are also working to design and build a stove that will get an Underwriters Laboratory (UL (safety organization)) listing, which would eliminate insurance issues. They even helped write the building code for Portland. In 2013, they were featured in a 4 DVD film series produced by Paul Wheaton. The films covered an entire workshop which they split into four sections: Fire Science, Sneaky Heat, Boom Squish, and Hot Rocket.[4] Additionally, Village Video produced a film with Ernie and Erica called How To Build Rocket Mass Heaters in which they again teach how to build and maintain these heaters. Adding to all of this, Erica and Ernie have written two books, *The Art of Fire and The Rocket Mass Heater Builder's Guide.*

7.3 See also

- Permaculture
- Ianto Evans
- Paul Wheaton
- Mike Oehler

7.4 References

[1] Mackay, Mary (November 10, 2012), "Stove Really Takes Off", *Times Colonist*, retrieved June 12, 2014

[2] Mackay, Mary (November 10, 2012), "Stove Really Takes Off", *Times Colonist*, retrieved June 12, 2014

[3] Grisak, Amy (June 12, 2013), "Rocket Mass Heating", *The New Pioneer*, retrieved June 12, 2014

[4] Wheaton, Paul (2013), *Rocket Stove Mass Heater*, retrieved June 12, 2014

Chapter 8

Folkewall

The **Folkewall**[1] is a construction with the dual functions of growing plants and purifying waste water. It was designed by Folke Günther[2] in Sweden.

Inspired by the "Sanitas wall" at Dr Gösta Nilsson's Sanitas farm project in Botswana, this technique makes an efficient use of space by fulfilling two essential functions: vertical plant growing and purification of greywater. This system is also known as a living wall or green wall.

8.1 Design

The basic design is a wall of hollow concrete slabs, with compartments opening on one or both sides of the wall. The hollows are filled with inert material like gravel, expanded clay aggregate, perlite, or vermiculite. It is designed to let the water trickle over the longest possible treatment path along the length of the wall among the pebbles.

The water is brought in at the top, and percolates following a zig-zag pattern inside the wall. As it does so the plant roots grow among the inert material and extract nutrients from the water. A film of beneficial bacteria grows over the pebbles, releasing the nutrients in the percolating greywater. At the bottom of the wall a container collects the purified water, which can then be used for non-potable household use, for watering the garden, or it can be returned to the top of the wall.[3]

8.2 Other considerations

Plants used: since the harvesting of the plants is a part of the purification process, fast growing, herbaceous crops are particularly suited for the Folkewall. Annual food crops are suitable, perennials like trees and shrubs should be avoided.

Greywater: The water feeding the plants in the wall must be free of heavy metals and/or unsafe pollutants, notably human waste. This requires using source-separating toilets.

8.3 Advantages

- **Better use of greywater**: most of the evaporation happens through the plant's leaves, which makes the method especially useful in arid climates. The Folkewall makes use of this aspect.

- **More efficient use of the area**. For example in greenhouses or other glazed areas where a wall is used as a greywater purification device, it also works as a **heat exchanger and buffer**.

- **Purification** of the percolating water, so greywater can be used as irrigation water.

- In warm climates, the wall can be used on the sunny side to **cool the building**.

- **Low-cost housing**: the combined use of Folkewalls and source-separating toilets would "reduce the infrastructure cost by about 30%".[4]

8.4 References

[1] Folkewall

[2] Folke Günther

[3] Living wall systems *(subscription required)*

[4] Growing walls

8.5 External links

- Environmental design of human settlements

- Engineered solutions for greywater purification and recycling

- Societal design for phosphorus recycling

- Botswana tries two dryland farming methods

Chapter 9

Food Not Lawns

Food Not Lawns is a de-centralized social movement focused on replacing urban lawns with food-producing organic gardens. The first group to use the name "Food Not Lawns" was founded in Eugene, Oregon in 1999[1] by Tobias Policha, Nick Routledge, and Heather Jo Flores. In 2006, Flores published the book *Food Not Lawns: How to Turn Your Yard into a Garden and Your Neighborhood into a Community*[2] A self-described "avant-gardening collective"[3] FNL's basic premise was to garner surplus resources, whether food, seeds, plants, tools, garden space, publications, or volunteer time, and channel them toward building better food security for the community at hand.

Born of Eugene's radical political organizing community in the late 1990s, Food Not Lawns evolved directly out of the local chapter of Food Not Bombs, a free food-sharing collective having a common concern with food justice issues, and a similar stewardship and democratic approach. Neither Food Not Lawns nor Food Not Bombs chapters answer to a central decision-making body. Rather, they are examples of movements that operate on a premise of anarchism, autonomy and self-organization.[4] Everyone is free to start a Food Not Lawns group.

Food Not Lawns chapters have common activities. They organize local seed swap events. They build community gardens. They generate publications,[5] whether on the web or in print. And they host work parties to help community members turn their lawns into gardens.

According to the Food Not Lawns International Website,[6] Food Not Lawns currently has more than 50 chapters worldwide. Similar movements have also started up, with names including Grow Food Not Lawns and Plant Food Not Lawns; merchandise can be found bearing any of the three slogans. What started as a small neighborhood garden project has become a central meme of the sustainability and permaculture movements.

9.1 References

[1] Dr. Susan Rubin, The Case Against Lawns

[2] Flores, Heather (2006). *Food Not Lawns, How to Turn Your Yard into a Garden and Your Neighborhood into a Community*. White River Junction, Vermont: Chelsea Green Publishing.

[3] Food Not Lawns original website

[4] Food Not Lawns interview from Oprah.com

[5] Copy of the FNL Weed Lover zine, June 2001

[6] Food Not Lawns International Website

Chapter 10

Grassed waterway

Grassed waterway in Velm, Belgium, during a sunny day

A **grassed waterway** consists in a 2-metre (6.6 ft) to 48-metre-wide (157 ft) native grassland strip of green belt. It is generally installed in the thalweg, the deepest continuous line along a valley or watercourse, of a cultivated dry valley in order to control erosion. A study carried out on a grassed waterway during 8 years in Bavaria showed that it can lead to several other types of positive impacts, e.g. on biodiversity.[1]

10.1   Distinctions

Confusion between "grassed waterway" and "vegetative filter strips" should be avoided. The latter are generally narrower (only a few metres wide) and rather installed along rivers as well as along or within cultivated fields. However, buffer strip can be a synonym, with shrubs and trees added to the plant component, as does a riparian zone.

Grassed waterway in Velm, Belgium, after a thunderstorm

10.2   Runoff and erosion mitigation

Runoff generated on cropland during storms or long winter rains concentrates in the thalweg where it can lead to rill or gully erosion.

Rills and gullies further concentrate runoff and speed up its transfer, which can worsen damage occurring downstream. This can result in a muddy flood.

In this context, a grassed waterway allows increasing soil cohesion and roughness. It also prevents the formation of rills and gullies. Furthermore, it can slow down runoff and allow its re-infiltration during long winter rains. In contrast, its infiltration capacity is generally not sufficient to reinfiltrate runoff produced by heavy spring and summer storms. It can therefore be useful to combine it with extra measures, like the installation of earthen dams across the grassed waterway, in order to buffer runoff temporarily.[2]

10.3 External links

- (Dutch) Water Agency of the Melsterbeek river, in Belgium (agency that had a pioneer role to implement erosion control measures in central Belgium)

10.4 References

[1] Fiener P., Auerswald K. (2003). Concept and effects of a multi-purpose grassed waterway. Soil Use and Management 19, 65-72.

[2] Evrard, O., Vandaele, K., van Wesemael, B., Bielders, C.L, 2008. A grassed waterway and earthen dams to control muddy floods from a cultivated catchment of the Belgian loess belt. Geomorphology 100, 419-428.

Chapter 11

Holzer Permaculture

The **Holzer Permaculture** is a branch of permaculture developed independently from the mainstream permaculture in Austria by Sepp Holzer. It is particularly noteworthy because it grew out of practical application and was relatively detached from the scientific community.

11.1 Intro

Sepp Holzer started reorganising his father's property according to ecological patterns in the early 1960s after he took over the farm. As an adolescent he conducted layman experiments with plants native to the area and learned from his own observations.

Since having taken over his father's property, he has expanded it from 24 to 45 hectares. according to his methods together with his wife.

His expanded farm now spans over 45 hectares of forest gardens, including 70 ponds, and is said to be the most consistent example of permaculture worldwide. In the past he has experimented with many different animals. As a result of these experiments, there is a huge role for animals in the Holzer Permaculture.

He has created some of the world's best examples of using ponds as reflectors to increase solar gain for passive solar heating of structures, and of using the microclimate created by rock outcrops to effectively change the hardiness zone for nearby plants. He has also done original work in the use of Hügelkultur and natural branch development instead of pruning (see Fruit tree pruning) to allow fruit trees to survive high altitudes and harsh winters.

11.2 Comparison to regular permaculture

It is difficult to make out differences between the methods and practices of Sepp Holzer in contrast to the more scientific and theoretical permacultural mainstream. Nevertheless, here are some major points to consider:

- His designs are mostly aimed at raising temperatures and creating micro-climates with rocks, ponds and living wind barriers, in an area with 4 °C on the average and −20 °C in the winter. The ponds he makes do not contain any pond liner. Instead, he makes the ponds watertight by sifting the fine from the coarse soil in the earth pile dug up with the excavator. The excavator is then used to pile only the coarse soil up into walls which are then tampered down using the excavator's bucket. The bottom of the pond, he makes watertight by vibrating the excavator's bucket when the pond has been filled with 30-40cm of water.

- Another aspect was the necessity of creating terraces on his farm's hillsides leading him to the use of heavy machinery (like excavators). Many of the terraces he construct are also given a humus storage ditch (placed in between the terraces).

- Famous is his Hügelkultur technique, which is basically the use of raised beds in which he uses bulky material such as tree trunks. On his farm, he made a pick-your-own area where visitors can pick their own produce from the raised beds and then pay for it at a counter upon leaving the area.

- He uses animal labor alongside human labor, working his farm with only two people. He optimizes the natural patterns of animal behavior to reduce human or machine-driven labor. As an example: he uses swine to "plow" new beds for sowing. This is a very effective way of digging, as the only thing he has to do is to throw some corn and fruit on the spot he wants dug up. A couple of days later, he can bring the pigs back to their enclosure and plant new plants in the bed. Holzer is able to successfully grow his plants without using any fertilizer. The animals he uses are all heirloom races, which are hardy and require (almost) no maintenance. Examples of the races of swine he uses are Mangalitza, Swabian Hall swine, Duroc, Turopolje, ... He also keeps quail, capercaillie, hazel grouse, wisent, scottish higland cattle, hungarian steppe cattle, Dahomey miniature cattle, American bison, yak, water buffalo. He has the animals live outside, in paddocks/shelters and has them share space with orchards and forests. Many of the fruit trees in the orchards (especially those on very sloped terrain) are used exclusively to feed the animals. They collect it as it falls to the ground (those trees are hence not being picked/ nor is produce of it sold)

- He also does not prune fruit trees much nor does he cut the lower branches on fruit trees (as this can hurt the tree -due to the lignification not being able to complete before frost and the fact that the unpruned fruit trees survives snow loads that will break pruned trees.[1]-. He also says that leaving the branches on protects the tree against browsing by animals. He does however makes a point of using deep-rooted pioneer plants such as lupins, sweet clover, lucerne and broom. These crops are said to aerate the soil and make sure no water is left standing near the tree. No use is made of wire meshes as protection against voles, since he states they are not efficient in preventing damage from voles using this technique anyway. He makes no use of contemporary fruit tree cultivars, but only uses (very strong/hardy) old, local (heirloom) cultivars. In addition to (old cultivars of) regular fruit trees, he also plants much fruit tree species that are specific to use in phytotherapy or which can only function as animal feed (i.e. crab apple, wild pear, wild cherry, blackthorn, rowan, wild service, service tree, cornelian cherry, snowy mespilus, ...). Similar to mainstream permaculture, he makes no use of chemical fertiliser or pesticides, at all.[2] It should be noted that due to the altitude he's at, his trees bear fruit later, meaning he can sell it after most (sealevel) farmers sold their produce. Due to this (and the fact that he produces heirloom fruit varieties, and not regular varieties), he is often able to get a better price for it. In some cases, customers (like distilleries) are even willing to pick the fruit themselves, eliminating the labour expense for him.

- He grows many old cereals on his farm, such as einkorn, emmer, black emmer, spelt, fichtelgebirgshafer, wild rye, black oats, naked oats, barley, Siberian grain (secale cereale, russian cultivar), tauernroggen, ...

- He also makes much use of green manure crops (like

stinging nettle, phacelia, yellow, white and narrow-leaved lupin, garden pea, grass pea, fodder & Kidney vetch, yellow, subterranean, Crimson, Persian, Egyptian, red & white sweet clover, Birdsfoot trefoil, lucerne, black medick, Sainfoin, Serradella, fiddleneck, sunflowers, Jerusalem artichoke, Gold-of-pleasure, ...)and grows these crops extensively on his farm. He leaves them standing in autumn, rather than digging them in. He instead relies on natural decay of the plants. He often relies on the natural spreading of the seeds of the crops for their re-sowing.

- Another aspect is the abandonment of other horticultural principles such as intercropping plants with very high and very low pH requirements (for example, Rhododendron with roses). Instead, Holzer mixes thirty or more different types of seeds in a bucket and tosses the mix richly onto a larger area.

11.3 The Krameterhof

Situated in Ramingstein on the slopes of Mount Schwarzenberg his farm (Krameterhof) lies at varying elevations ranging from 1100 to 1500 metres above sea level.

The exceptionally harsh climatic conditions in the area are generally considered inappropriate for farming. Nevertheless, he has managed to cultivate a variety of crops and even exotic plants like Kiwis and Sweet Chestnut.

The Krameterhof is less an operational enterprise, in terms of crop-yield (although it does provide numerous sorts of produce for the community), and more a fully functional showcase or research station for permaculture.

Endangered livestock species and rare alpine- and cultural plant species are integrated into the farm.

11.4 Publications translated into English

Most of Holzer's books are published in German through Leopold Stocker Verlag, an Austrian publisher based in Graz.

- "The Rebel Farmer"

- "Sepp Holzer's Permaculture"

11.5 References

[1] Ferguson, Julia. Ecological farming, permaculture; Alpine Garden of Eden proves Mother Nature knows best. Reuters. Retrieved August 7, 2011.

[2] Sepp Holzer's permaculture by Sepp Holzer]

11.6 External links

- Holzer Permaculture
- Video "Farming With Nature"
- Sepp Holzer's Permaculture

Chapter 12

Intercropping

Intercropping is a multiple cropping practice involving growing two or more crops in proximity. The most common goal of intercropping is to produce a greater yield on a given piece of land by making use of resources that would otherwise not be utilized by a single crop.[1] Careful planning is required, taking into account the soil, climate, crops, and varieties. It is particularly important not to have crops competing with each other for physical space, nutrients, water, or sunlight. Examples of intercropping strategies are planting a deep-rooted crop with a shallow-rooted crop, or planting a tall crop with a shorter crop that requires partial shade. Inga alley cropping has been proposed as an alternative to the ecological destruction of slash-and-burn farming.[2]

When crops are carefully selected, other agronomic benefits are also achieved. Lodging-prone plants, those that are prone to tip over in wind or heavy rain, may be given structural support by their companion crop.[3] Creepers can also benefit from structural support. Some plants are used to suppress weeds or provide nutrients.[4] Delicate or light-sensitive plants may be given shade or protection, or otherwise wasted space can be utilized. An example is the tropical multi-tier system where coconut occupies the upper tier, banana the middle tier, and pineapple, ginger, or leguminous fodder, medicinal or aromatic plants occupy the lowest tier.

Intercropping of compatible plants also encourages biodiversity, by providing a habitat for a variety of insects and soil organisms that would not be present in a single-crop environment. This in turn can help limit outbreaks of crop pests by increasing predator biodiversity.[5] Additionally, reducing the homogeneity of the crop increases the barriers against biological dispersal of pest organisms through the crop.

There are several ways pests can be controlled through intercropping:

- **Trap cropping,** this involves planting a crop nearby that is more attractive for pests compared to the production crop, the pests will target this crop and not the production crop.

- **Repellant intercrops,** an intercrop that has a repellent effect to certain pests can be used. This system involved the repellant crop masking the smell of the production crop in order to keep pests away from it.

- **Push-pull cropping,** this is a mixture of trap cropping and repellant intercropping. An attractant crop attracts the pest and a repellant crop is also used to repel the pest away. [6]

The degree of spatial and temporal overlap in the two crops can vary somewhat, but both requirements must be met for a cropping system to be an intercrop. Numerous types of intercropping, all of which vary the temporal and spatial mixture to some degree, have been identified.[7][8] These are some of the more significant types:

- **Mixed intercropping**, as the name implies, is the most basic form in which the component crops are totally mixed in the available space.

- **Row cropping** involves the component crops arranged in alternate rows. Variations include alley cropping, where crops are grown in between rows of trees, and strip cropping, where multiple rows, or a strip, of one crop are alternated with multiple rows of another crop. A new version of this is to intercrop rows of solar photovoltaic modules with agriculture crops. This practice is called agrivoltaics.[9]

- **Temporal intercropping** uses the practice of sowing a fast-growing crop with a slow-growing crop, so that the fast-growing crop is harvested before the slow-growing crop starts to mature.

- Further temporal separation is found in **relay cropping**, where the second crop is sown during the growth, often near the onset of reproductive development or fruiting, of the first crop, so that the first crop is harvested to make room for the full development of the second.

12.1 See also

- Agrivoltaics

- Allotment garden

- Asset-based community development (ABCD)

- Community Food Security Coalition

- Community gardening

- Container garden

- Companion planting

- Ecological sanitation

- Food-feed system

- Forest gardening

- Gardening

- Green wall

- Monoculture

- Organic farming

- Permaculture

- Sustainable agriculture

12.2 References

[1] Ouma, George; Jeruto, P (2010). "Sustainable horticultural crop production through intercropping: The case of fruits and vegetable crops: A review" (PDF). *Agriculture and Biology Journal of North America*. **1** (5): 1098–1105.

[2] Elkan, Daniel. *Slash-and-burn farming has become a major threat to the world's rainforest* The Guardian 21 April 2004

[3] Trenbath, B.R. 1976. Plant interactions in mixed cropping communities. pp. 129–169 in R.I. Papendick, A. Sanchez, G.B. Triplett (Eds.), *Multiple Cropping*. ASA Special Publication 27. American Society of Agronomy, Madison, WI.

[4] Mt. Pleasant, Jane (2006). "The science behind the Three Sisters mound system: An agronomic assessment of an indigenous agricultural system in the northeast". In John E. Staller; Robert H. Tykot; Bruce F. Benz. *Histories of maize: Multidisciplinary approaches to the prehistory, linguistics, biogeography, domestication, and evolution of maize*. Amsterdam. pp. 529–537.

[5] Miguel Angel Altieri; Clara Ines Nicholls (2004). *Biodiversity and Pest Management in Agroecosystems, Second Edition*. Psychology Press.

[6] "Controlling Pests with Plants: The power of intercropping - UVM Food Feed". *UVM Food Feed*. 2014-01-09. Retrieved 2016-12-01.

[7] Andrews, D.J., A.H. Kassam. 1976. The importance of multiple cropping in increasing world food supplies. pp. 1–10 in R.I. Papendick, A. Sanchez, G.B. Triplett (Eds.), *Multiple Cropping*. ASA Special Publication 27. American Society of Agronomy, Madison, WI.

[8] Lithourgidis, A.S.; Dordas, C.A.; Damalas, C.A.; Vlachostergios, D.N. (2011). "Annual intercrops: an alternative pathway for sustainable agriculture" (PDF). *Australian Journal of Crop Science*. **5** (4): 396–410.

[9] Dinesh, Harshavardhan; Pearce, Joshua M. (2016-02-01). "The potential of agrivoltaic systems". *Renewable and Sustainable Energy Reviews*. **54**: 299–308. doi:10.1016/j.rser.2015.10.024.

[10] *Improving nutrition through home gardening*, Home Garden Technology Leaflet 13: Multilayer cropping, FAO, 2001

12.3 External links

- Intercropping at Washington State University

Chapter 13

Keyline design

For the graphics design term, see Keyline.

Keyline design is a technique for maximizing beneficial

A keyline irrigation channel

use of water resources of a piece of land. The *Keyline* refers to a specific topographic feature linked to water flow. Beyond that however, Keyline can be seen as a collection of design principles, techniques and systems for development of rural and urban landscapes.

Keyline design was developed in Australia by farmer and engineer P. A. Yeomans, and described and explained in his books *The Keyline Plan*, *The Challenge of Landscape*, *Water For Every Farm*, and *The City Forest*.

13.1 Application

P. A. Yeomans published the first book on Keyline in 1954. Yeomans described a system of amplified contour ripping to control rainfall runoff and enable fast flood irrigation of undulating land without the need for terracing.

Keyline designs include irrigation dams equipped with through-the-wall lockpipe systems to gravity feed irrigation, stock water, and yard water. Graded earth channels may be interlinked to broaden the catchment areas of high dams, conserve the height of water, and transfer rainfall runoff into the most efficient high dam sites. Roads follow both ridge lines and water channels to provide easier movement across the land.[1]

13.2 Keyline Scale of Permanence

The backbone of Yeomans' keyline design system, the outcome of fifteen years of adaptive experimentation, is Yeomans' Keyline Scale of Permanence (KSOP), which identifies typical farms' elements ordered according to their degree of permanence:

- Climate

- Landshape

- Water Supply

- Roads/Access

- Trees

- Structures

- Subdivision Fences

- Soil

Keyline considers these elements when planning the placement of water storage, roads, trees, buildings and fences.

On undulating land, a keyline approach involves identifying several features namely ridges and valleys and the natural water courses seeking optimum water storage sites and potential interconnecting channels.

The water lines identified from the land-form subsequently provide optimal locations for the various less permanent elements (roads, fences, trees, and buildings) to optimize the natural potential of the landscape.

Rancho San Ricardo, México

13.3   Keypoint

In a smooth grassy valley, a location called the **keypoint** can be found where the lower and flatter portion of a primary valley floor suddenly steepens. The **keyline** of this primary valley is revealed by pegging a contour line through the keypoint, within the valley shape. All the points on the line are at the same elevation as the keypoint. Contour plowing parallel to the Keyline, both above and below will automatically become "off-contour" but the developing pattern will tend to drift rainwater runoff away from the valley centre and incidentally, prevent erosion.

Keyline pattern cultivation on ridge shapes is done parallel to any suitable contour but only working on the upper side of the contour guide line. This automatically develops a pattern of off-contour cultivation in which all the rip marks left in the soil will slope down towards the centre of the ridge shape. This pattern of cultivation allows more time for water to soak in. Keyline pattern cultivation also enables controlled flood irrigation of undulating land, which further assists in the fast development of deep biologically fertile soil, which results in improving soil nutrition and health.

In many countries, including Australia, it is important to get optimum absorption of rainfall and keyline cultivation does this as well as delaying the potentially damaging concen-

tration of runoff. Yeomans' technique differs from traditional contour plowing in several important respects. Random contour plowing also becomes off contour but usually with the opposite effect on runoff water causing it to quickly shed off ridge shapes and be concentrated in valleys. The limitations of the traditional system of **soil conservation**, with its "safe disposal" approach to farm water was an important motivational factor in the development of the keyline system.

13.4   Applications

David Holmgren, one of the founders of Permaculture, used Yeoman's keyline principle extensively in the formulation of Permaculture concepts and the design of sustainable human settlements and organic farms.

Darren J. Doherty has extensive experience across the world in keyline project design, development, management & training.

A topographical example can be seen on (37°09′33″S 144°15′08″E / 37.159154°S 144.252248°E[2]).

Keyline also includes concepts for rapid soil fertility enhancement and these concepts are explored in *Priority One* by P. A. Yeomans' son Allan. Yeomans and his sons were also instrumental in the design and production of special plows and cultivating equipment for use in conjunction with the keyline methodology.

13.5   See also

13.6   References

13.6.1   Notations

- Yeomans, P.A. (1954). *The Keyline Plan* (Free online). OCLC 21106239.

- Yeomans, P.A. (1958). *The Challenge of Landscape : the development and practice of keyline* (Free online). Sydney NSW: Keyline. OCLC 10466838.

- Yeomans, P.A. (1973). *Water for Every Farm: A practical irrigation plan for every Australian property*. Sydney NSW: K.G. Murray. ISBN 0-646-12954-6. ISBN 0-909325-29-4.

- Yeomans, P.A. (1971). *The City Forest* (Free online). Keyline. ISBN 0-9599578-0-4. OCLC 515050.

- Yeomans, P.A.; Yeomans, K.B. (1993). *Water for Every Farm — Yeomans Keyline Plan*. Keyline Designs. ISBN 0646129546. 2002 ISBN 0646418750

- Yeomans, P.A.; Yeomans, K.B. (2008). *Water for Every Farm — Yeomans Keyline Plan*. Keyline Designs. ISBN 1438225784. External link in |publisher= (help)

- Yeomans, A. (2005). *Priority One: Together we Can Beat Global Warming* (Online). Keyline. ISBN 0-646-43805-0.

- MacDonald-Holmes, J. "Geographical and Topographical Basis of Keyline". Archived from the original on August 15, 2010.

- Spencer, L (2006). "Keyline and Fertile Futures". Archived from the original on December 9, 2009.

13.6.2 Footnotes

[1] Keyline Designs website

[2] Fryers Forest on WikiMaps

13.7 External links

Chapter 14

Leaf mold

Leaf mold (**Leaf mould** outside of America) is the product of slow decomposition of deciduous shrub and tree leaves. It is a form of compost produced primarily by fungal breakdown[1]

14.1 Description

Leaves shed in autumn tend to have a very low nitrogen content and are often dry. Their main constituent is cellulose and lignin.[2] Because of the this, autumn leaves break down far more slowly than most other compost ingredients with very little bacterial decomposition involved.

14.2 Time and process

Fungal decomposition of a heap of leaves in the open can can take between one and two years to break down into a dark brown fine powdery humic matter. During the two to three years that the process takes to complete in damp temperate climates , a succession of different fungal species may be involved.[3] A range of micro fauana are also involved in converting the leaf material into a fine grained humus and these include many isopods, millipedes,, earthworms etc.

14.3 Uses

In the natural environment the slow decomposition of leaves provides a moist growing medium for young plants and also protects the ground from drying out during periods of low rainfall. It is a significant component of soil organic matter, particularly in temperate deciduous woodland. The slow rate of decomposition allows the plant nutrients bound up in the leaves to be released slowly back into the environment where they can be re-used by plants. Autumn leaves are often collected as part of gardening or farming and kept in pits or containers so that the leaf mold can be used in the garden. The presence of oxygen from the air and sufficient moisture are essential for leaf decomposition. Leaf mold is not high in nutrient content, but it is excellent humic soil conditioner because of its ability to retain moisture and provide a good growing medium for seedling roots. Leaves collected off roads and pavements may be contaminated by pollutants which can become more concentrated as the leaves decompose into a smaller volume [4]

14.4 See also

- Worm compost
- Spent mushroom compost
- Recycling

14.5 References

[1] Compost organisms

[2] "Compost Chemistry". Cornell University. Retrieved 6 October 2016.

[3] Jana Voříšková1; Petr Baldrian (11 October 2012). "Fungal community on decomposing leaf litter undergoes rapid successional changes". The ISME Journal. Retrieved 7 October 2016.

[4] "Leaf litter in street sweepings: investigation into collection and treatment" (PDF). The Environment Agency. Retrieved 6 October 2016.

14.6 External links

- BBC Gardening How to Make Leaf Mould
- Green Fingers How to Make Leaf Mould
- Leaves & Leaf Mold, nature's mulch & top-coating at The Garden of Paghat

Chapter 15

Mulch

A **mulch** is a layer of material applied to the surface of an area of soil. Its purpose is any or all of the following:

- to conserve moisture

- to improve the fertility and health of the soil

- to reduce weed growth

- to enhance the visual appeal of the area

A mulch is usually but not exclusively organic in nature. It may be permanent (e.g. plastic sheeting) or temporary (e.g. bark chips). It may be applied to bare soil, or around existing plants. Mulches of manure or compost will be incorporated naturally into the soil by the activity of worms and other organisms. The process is used both in commercial crop production and in gardening, and when applied correctly can dramatically improve soil productivity.[1]

15.1 Uses

Many materials are used as mulches, which are used to retain soil moisture, regulate soil temperature, suppress weed growth, and for aesthetics.[2] They are applied to the soil surface,[3] around trees, paths, flower beds, to prevent soil erosion on slopes, and in production areas for flower and vegetable crops. Mulch layers are normally two inches or more deep when applied.[4][5]

They are applied at various times of the year depending on the purpose. Towards the beginning of the growing season mulches serve initially to warm the soil by helping it retain heat which is lost during the night. This allows early seeding and transplanting of certain crops, and encourages faster growth. As the season progresses, mulch stabilizes the soil temperature and moisture, and prevents the growing of weeds from seeds.[6] In temperate climates, the effect of mulch is dependent upon the time of year they are applied and when applied in fall and winter, are used to delay the growth of perennial plants in the spring or prevent growth in winter during warm spells, which limits freeze thaw damage.[7]

The effect of mulch upon soil moisture content is complex. Mulch forms a layer between the soil and the atmosphere which prevents sunlight from reaching the soil surface, thus reducing evaporation. However, mulch can also prevent water from reaching the soil by absorbing or blocking water from light rains.

In order to maximise the benefits of mulch, while minimizing its negative influences, it is often applied in late spring/early summer when soil temperatures have risen sufficiently, but soil moisture content is still relatively high.[8] However, permanent mulch is also widely used and valued for its simplicity, as popularized by author Ruth Stout, who said, "My way is simply to keep a thick mulch of any vegetable matter that rots on both sides of my vegetable and flower garden all year long. As it decays and enriches the soils, I add more."[9]

Plastic mulch used in large-scale commercial production is laid down with a tractor-drawn or standalone layer of plastic mulch. This is usually part of a sophisticated mechanical process, where raised beds are formed, plastic is rolled out on top, and seedlings are transplanted through it. Drip irrigation is often required, with drip tape laid under the plastic, as plastic mulch is impermeable to water.

15.2 Materials

Materials used as mulches vary and depend on a number of factors. Use takes into consideration availability, cost, appearance, the effect it has on the soil—including chemical reactions and pH, durability, combustibility, rate of decomposition, how clean it is—some can contain weed seeds or plant pathogens.[6]

A variety of materials are used as mulch:

- Organic residues: grass clippings, leaves, hay, straw, kitchen scraps comfrey, shredded bark, whole

Rubber mulch nuggets in a playground. The white fibers are nylon cords, which are present in the tires from which the mulch is made.

Shredded wood used as mulch. This type of mulch is often dyed to improve its appearance in the landscape.

Pine needles used as mulch. Also called "pinestraw" in the southern US.

bark nuggets, sawdust, shells, woodchips, shredded newspaper, cardboard, wool, animal manure, etc.

Aged Compost mulch on a flower bed

Crushed stone mulch

Spring daffodils push through shredded wood mulch

Many of these materials also act as a direct composting system, such as the mulched clippings of a mulching lawn mower, or other organics applied as sheet composting.

- Compost: fully composted materials are used to avoid possible phytotoxicity problems. Materials that are free of seeds are ideally used, to prevent weeds being introduced by the mulch.

- Old carpet (synthetic or natural): makes a free, readily available mulch.[10]

- Rubber mulch: made from recycled tire rubber.

- Plastic mulch: crops grow through slits or holes in thin plastic sheeting. This method is predominant in large-scale vegetable growing, with millions of acres cultivated under plastic mulch worldwide each year (disposal of plastic mulch is cited as an environmental problem).

- Rock and gravel can also be used as a mulch. In cooler climates the heat retained by rocks may extend the growing season.

In some areas of the United States, such as central Pennsylvania and northern California, mulch is often referred to as "tanbark", even by manufacturers and distributors. In these areas, the word "mulch" is used specifically to refer to very fine tanbark or peat moss.

15.2.1 Organic mulches

Organic mulches decay over time and are temporary. The way a particular organic mulch decomposes and reacts to wetting by rain and dew affects its usefulness.

Some mulches such as straw, peat, sawdust and other wood products may for a while negatively affect plant growth because of their wide carbon to nitrogen ratio,[11] because bacteria and fungi that decompose the materials remove nitrogen from the surrounding soil for growth.[12][13] However, whether this effect has any practical impact on gardens is disputed by researchers and the experience of gardeners.[14] Organic mulches can mat down, forming a barrier that blocks water and air flow between the soil and the atmosphere. Vertically applied organic mulches can wick water from the soil to the surface, which can dry out the soil.[15] Mulch made with wood can contain or feed termites, so care must be taken about not placing mulch too close to houses or building that can be damaged by those insects. Some mulch manufacturers recommend putting mulch several inches away from buildings.

Commonly available organic mulches include:[6]

Leaves

- *Leaves* from deciduous trees, which drop their foliage in the autumn/fall. They tend to be dry and blow around in the wind, so are often chopped or shredded before application. As they decompose they adhere to each other but also allow water and moisture to seep down to the soil surface. Thick layers of entire leaves, especially of maples and oaks, can form a soggy mat in winter and spring which can impede the new growth lawn grass and other plants. Dry leaves are used as winter mulches to protect plants from freezing and thawing in areas with cold winters, they are normally removed during spring.

Grass clippings

- *Grass clippings*, from mowed lawns are sometimes collected and used elsewhere as mulch. Grass clippings are dense and tend to mat down, so are mixed with tree leaves or rough compost to provide aeration and to facilitate their decomposition without smelly putrefaction. Rotting fresh grass clippings can damage plants; their rotting often produces a damaging buildup of trapped heat. Grass clippings are often dried thoroughly before application, which mediates against rapid decomposition and excessive heat generation. Fresh green grass clippings are relatively high in nitrate content, and when used as a mulch, much of the nitrate is returned to the soil, but the routine removal of grass clippings from the lawn results in nitrogen deficiency for the lawn.

Peat moss

- *Peat moss*, or sphagnum peat, is long lasting and packaged, making it convenient and popular as a mulch. When wetted and dried, it can form a dense crust that does not allow water to soak in. When dry it can also burn, producing a smoldering fire. It is sometimes mixed with pine needles to produce a mulch that is friable. It can also lower the pH of the soil surface, making it useful as a mulch under acid loving plants.

However peat bogs are a valuable wildlife habitat, and peat is also one of the largest stores of carbon (in Britain, out of a total estimated 9952 million tonnes of carbon in British vegetation and soils, 6948 million tonnes carbon are estimated to be in Scottish, mostly peatland, soils[16]), so gardeners who wish to protect the environment will choose more sustainable alternatives.[17]

Wood chips

- *Wood chips* are a byproduct of the pruning of trees by arborists, utilities and parks; they are used to dispose of bulky waste. Tree branches and large stems are rather coarse after chipping and tend to be used as a mulch at least three inches thick. The chips are used to conserve soil moisture, moderate soil temperature and suppress weed growth. The decay of freshly produced chips from recently living woody plants, consumes nitrate; this is often off set with a light application of a high-nitrate fertilizer. Wood chips are most often used under trees and shrubs. When used around soft stemmed plants, an unmulched zone is left around the plant stems to prevent stem rot or other possible diseases. They are often used to mulch trails, because they are readily produced with little additional cost outside of the normal disposal cost of tree maintenance. Wood chips come in various colors.

Woodchip mulch

- *Woodchip mulch* is a byproduct of reprocessing used (untreated) timber (usually packaging pallets), to dispose of wood waste by creating woodchip mulch. The chips are used to conserve soil moisture, moderate soil temperature and suppress weed growth. Woodchip mulch is often used under trees, shrubs or large planting areas and can last much longer than arborist mulch. In addition, many consider woodchip mulch to be visually appealing, as it comes in various colors. Woodchips can also be reprocessed into playground woodchip to be used as an impact-attenuating playground surfacing.

Bark chips

Bark chips

- *Bark chips* of various grades are produced from the outer corky bark layer of timber trees. Sizes vary from thin shredded strands to large coarse blocks. The finer types are very attractive but have a large exposed surface area that leads to quicker decay. Layers two or

three inches deep are usually used, bark is relativity inert and its decay does not demand soil nitrates. Bark chips are also available in various colors.

Straw mulch / field hay / salt hay

Permaculture garden with a fruit tree, herbs, flowers and vegetables mulched with hay

- *Straw mulch* or *field hay* or *salt hay* are lightweight and normally sold in compressed bales. They have an unkempt look and are used in vegetable gardens and as a winter covering. They are biodegradable and neutral in pH. They have good moisture retention and weed controlling properties but also are more likely to be contaminated with weed seeds. Salt hay is less likely to have weed seeds than field hay. Straw mulch is also available in various colors.

Cardboard / newspaper

- *Cardboard* or *newspaper* can be used as mulches. These are best used as a base layer upon which a heavier mulch such as compost is placed to prevent the lighter cardboard/newspaper layer from blowing away. By incorporating a layer of cardboard/newspaper into a mulch, the quantity of heavier mulch can be reduced, whilst improving the weed suppressant and moisture retaining properties of the mulch.[8] However, additional labour is expended when planting through a mulch containing a cardboard/newspaper layer, as holes must be cut for each plant. Sowing seed through mulches containing a cardboard/newspaper layer is impractical. Application of newspaper mulch in windy weather can be facilitated by briefly pre-soaking the newspaper in water to increase its weight.

Carpet

- Synthetic carpet that is composed of artificial fibers may be removed after planting to prevent fibers taking a long time to decompose, whereas carpet made from natural fibers may be kept in place, blocking competition from weeds. Rain is absorbed by carpet and then slowly released into the soil, reducing watering needs.[18]

15.3 Colored Mulch

Some organic mulches are colored red, brown, black, and other colors. Isopropanolamine, specifically 1-Amino-2-propanol or DOW™ monoisopropanolamine, may be used as a pigment dispersant and color fastener in these mulches.[19][20][21][22] Types of mulch which can be dyed include: wood chips, bark chips (barkdust) and pine straw. Colored mulch is made by dyeing the mulch in a water-based solution of colorant and chemical binder. When colored mulch first entered the market, most formulas were suspected to contain toxic, heavy metals and other contaminates. Today, "current investigations indicate that mulch colorants pose no threat to people, pets or the environment. The dyes currently used by the mulch and soil industry are similar to those used in the cosmetic and other manufacturing industries (i.e., iron oxide)," as stated by the Mulch and Soil Council.[23] Colored mulch can be applied anywhere non-colored mulch is used (such as large bedded areas or around plants) and features many of the same gardening benefits as traditional mulch, such as improving soil productivity and retaining moisture.[24] As mulch decomposes, just as with non-colored mulch, more mulch may need to be added to continue providing benefits to the soil and plants. However, if mulch is faded, spraying dye to previously spread mulch in order to restore color is an option.[25]

15.4 Anaerobic (sour) mulch

Mulch normally smells like freshly cut wood, but sometimes develops a toxicity that causes it to smell like vinegar, ammonia, sulfur or silage. This happens when material with ample nitrogen content is not rotated often enough and it forms pockets of increased decomposition. When this occurs, the process may become anaerobic and produce these phytotoxic materials in small quantities. Once exposed to the air, the process quickly reverts to an aerobic process, but these toxic materials may be present for a period of time. If the mulch is placed around plants before the toxicity has had a chance to dissipate, then the plants could very likely be damaged or killed depending on their hardiness. Plants that are predominantly low to the ground or freshly planted are the most susceptible, and the phytotoxicity may prevent germination of some seeds.[26]

If sour mulch is applied and there is plant kill, the best thing to do is to water the mulch heavily. Water dissipates the chemicals faster and refreshes the plants. Removing the offending mulch may have little effect, because by the time plant kill is noticed, most of the toxicity is already dissipated. While testing after plant kill will not likely turn up anything, a simple pH check may reveal high acidity, in the range of 3.8 to 5.6 instead of the normal range of 6.0 to 7.2. Finally, placing a bit of the offending mulch around another plant to check for plant kill will verify if the toxicity has departed. If the new plant is also killed, then sour mulch is probably not the problem.

15.5 Groundcovers (living mulches)

Main articles: Groundcovers and Living mulch

Groundcovers are plants which grow close to the ground, under the main crop, to slow the development of weeds and provide other benefits of mulch. They are usually fast-growing plants that continue growing with the main crops. By contrast, cover crops are incorporated into the soil or killed with herbicides. However, live mulches also may need to be mechanically or chemically killed eventually to prevent competition with the main crop.[27]

Some groundcovers can perform additional roles in the garden such as nitrogen fixation in the case of clovers, dynamic accumulation of nutrients from the subsoil in the case of creeping comfrey (*Symphytum ibericum*), and even food production in the case of *Rubus tricolor*.[28]

15.6 On-site mulch production

Owing to the great bulk of mulch which is often required on a site, it is often impractical and expensive to source and import sufficient mulch materials. An alternative to importing mulch materials is to grow them on site in a "mulch garden" - an area of the site dedicated entirely to the production of mulch which is then transferred to the growing area.[28] Mulch gardens should be sited as close as possible to the growing area so as to facilitate transfer of mulch materials.[28]

15.7 Mulching (composting) over unwanted plants

Main article: Sheet mulching

Sufficient mulch over plants will destroy them, and may be more advantageous than using herbicide, cutting, mowing, pulling, raking, or tilling. The higher the temperature that this "mulch" is composted, the quicker the reduction of undesirable materials. "Undesirable materials" may include living seed, plant "trash", as well as pathogens such as from animal feces, urine (e.g. hantavirus), fleas, lice, ticks, etc.

In some ways this improves the soil by attracting and feeding earthworms, and adding humus. Earthworms "till" the soil, and their feces are among the best fertilizers and soil conditioners.

Urine may be toxic to plants if applied to growing areas undiluted. See Compost ingredients: Human Waste.

15.8 Polypropylene and polyethylene mulch

Polypropylene mulch is made up of polypropylene polymers where polyethylene mulch is made up of polyethylene polymers. These mulches are commonly used in many plastics. Polyethylene is used mainly for weed reduction, where polypropylene is used mainly on perennials.[29] This mulch is placed on top of the soil and can be done by machine or hand with pegs to keep the mulch tight against the soil. This mulch can prevent soil erosion, reduce weeding, conserve soil moisture, and increase temperature of the soil.[30] Ultimately this can reduce the amount of work a farmer may have to do, and the amount of herbicides applied during the growing period. The black and clear mulches capture sunlight and warm the soil increasing the growth rate. White and other reflective colours will also warm the soil, but they do not suppress weeds as well.[30] This mulch may require other sources of obtaining water such as drip irrigation since it can reduce the amount of water that reaches the soil.[30] This mulch needs to be manually removed at the end of the season since when it starts to break down it breaks down into smaller pieces.[31] If the mulch is not removed before it starts to break down eventually it will break down into ketones and aldehydes polluting the soil.[31] This mulch is technically biodegradable but does not break down into the same materials the more natural biodegradable mulch does.

15.9 Biodegradable mulch

Quality biodegradable mulches are made out of plant starches and sugars or polyester fibres. These starches can come from plants such as wheat and corn.[32] These mulch films may be a bit more permeable allowing more water into the soil. This mulch can prevent soil erosion, reduce weeding, conserve soil moisture, and increase temperature of the soil.[30] Ultimately this can reduce the amount of herbicides used and manual labour farmers may have to do throughout the growing season. At the end of the season these mulches will start to break down from heat. Microorganisms in the soil break down the mulch into two components, water and CO_2, leaving no toxic residues behind.[32] This source of mulch is even less manual labour since it does not need to be removed at the end of the season and can actually be tilled into the soil.[32] With this mulch its important to take into consideration that its mulch is more delicate then other kinds. It should be placed on a day which is not too hot and with less tension then other synthetic mulches.[32] These also can be placed by machine or hand but its ideal to have a more starchy mulch that will allow it to stick to the soil better.

15.10 See also

- Forestry mulching
- Good Agricultural Practices
- Rubber mulch
- Plasticulture
- Integrated pest management
- Living mulch

15.11 References

[1] *RHS A-Z encyclopedia of garden plants*. United Kingdom: Dorling Kindersley. 2008. p. 1136. ISBN 1405332964.

[2] Alfred J. Turgeon; Lambert Blanchard McCarty; Nick Edward Christians (2009). *Weed control in turf and ornamentals*. Prentice Hall. p. 126. ISBN 978-0-13-159122-6.

[3] Mahesh K. Upadhyaya; Robert E. Blackshaw (2007). *Nonchemical Weed Management: Principles, Concepts and Technology*. CABI. pp. 135–. ISBN 978-1-84593-291-6.

[4] *Vegetable Gardening: Growing and Harvesting Vegetables*. Murdoch Books. 2004. pp. 110–. ISBN 978-1-74045-519-0.

[5] Dennis R. Pittenger (2002). *California Master Gardener Handbook*. UCANR Publications. pp. 567–. ISBN 978-1-879906-54-9.

[6] Louise; Bush-Brown, James (1996). *America's garden book*. New York: Macmillan USA. p. 768. ISBN 0-02-860995-6{{inconsistent citations}}

[7] Leon C. Snyder (2000). *Gardening in the Upper Midwest*. University of Minnesota Press. pp. 47–. ISBN 978-0-8166-3838-3.

[8] Patrick Whitefield, 2004, *The Earth Care Manual*, Permanent Publications, ISBN 978-1-85623-021-6

[9] Stout, Ruth. *Gardening Without Work*. Devon-Adair Press, 1961. Reprinted by Norton Creek Press, 2011, pp. 6-7. ISBN 978-0-9819284-6-3

[10] Galloway, David. "Get Your New Garden Ready for Spring with Old Carpet". *Lifehacker*. Retrieved 2016-02-19.

[11] http://www.eau.ee/~{}agronomy/vol07Spec1/p7sI53.pdf

[12] http://joa.isa-arbor.com/request.asp?JournalID=1&ArticleID=3111&Type=2

[13] Jeff Gillman (1 February 2008). *The Truth About Organic Gardening: Benefits, Drawbacks, and the Bottom Line*. Timber Press. pp. 51–. ISBN 978-1-60469-005-7.

[14] Stout, Ruth. *Gardening Without Work*. Devon-Adair Press, 1961. Reprinted by Norton Creek Press, 2011, pp. 192-193. ISBN 978-0-9819284-6-3

[15] David A. Bainbridge (11 June 2007). *A Guide for Desert and Dryland Restoration: New Hope for Arid Lands*. Island Press. pp. 239–. ISBN 978-1-61091-082-8.

[16] Milne, R.; T. A. Brown (1997). "Carbon in the vegetation and soils of Great Britain". *Journal of Environmental Management*. **49**: 413–433. doi:10.1006/jema.1995.0118.

[17] Walker, John (2011). *How to Create an Eco Garden: The practical guide to greener, planet-friendly gardening*. Wigston, Leicestershire: Aquamarine. p. 33. ISBN 9781903141892.

[18] Galloway, David. "Get Your New Garden Ready for Spring with Old Carpet". *Lifehacker*. Retrieved 2016-02-19.

[19] Product Information - DOW™ Monoisopropanolamine (MIPA)

[20] Product Safety Assessment - DOW™ Monoisopropanolamine

[21] 2010 Mulch Magic Red Material Safety Data Sheet

[22] 2007 Mulch Magic Red Material Safety Data Sheet

[23] http://www.mulchandsoilcouncil.org/faqs/mulch.php

[24] http://www.hort.purdue.edu/ext/mulch.html

[25] http://homeguides.sfgate.com/there-spray-can-use-renew-mulch-color-64513.html

[26] Beware of Sour Mulch

[27] Brandsaeter et al. 1998, Tharp and Kells, 2001

[28] Jacke and Toensmeier, Edible Forest Gardening, vol. II

[29] Dovorak, P. "BLACK POLYPROPYLENE MULCH TEXTILE IN ORGANIC AGRICULTURE" (PDF). *Czech University of Life Science Prague, Kamýcká*. **52**. Retrieved 16 November 2014.

[30] Shonbeck, Dr. Mark (12 September 2012). "Synthetic Mulching Materials for Weed Management". *Extension*. Retrieved 16 November 2014.

[31] Corbin, A (2013). "Using Biodegradable Plastics as Agricultural Mulches." (PDF). Retrieved 16 November 2014.

[32] "Biodegradable Mulch" (PDF). *Penn State Extension*. Retrieved 16 November 2014.

15.12 External links

- Mulching Trees & Shrubs

Chapter 16

Mycoforestry

Amanita *species are ectomycorrhizal with many trees*

Mycoforestry is an ecological forest management system implemented to enhance forest ecosystems and plant communities through the introduction of mycorrhizal and saprotrophic fungi. Mycoforestry is considered a type of permaculture[1] and can be implemented as a beneficial component of an agroforestry system. Mycoforestry can enhance the yields of tree crops and produce edible mushrooms, an economically valuable product. By integrating plant-fungal associations into a forestry management system, native forests can be preserved, wood waste can be recycled back into the ecosystem, planted restoration sites are enhanced, and the sustainability of forest ecosystems are improved.[2] Mycoforestry is an alternative to the practice of clearcutting, which removes dead wood from forests, thereby diminishing nutrient availability and reducing soil depth.[3]

16.1   Selection of fungal species

According to Paul Stamets, the first principle for the creation of a mycoforestry system is to utilize native fungal species. Implementing a mycoforestry system provides the potential of improving restoration efforts and the possibility of economic gain through mushroom cropping and harvesting. However to utilize native fungal flora, first the relationships between present fungal species and growth substrate, and habitat need to be studied.

A simple way to introduce a mycoforestry system and enhance out-plantings for crops and forest restoration sites is to "use mycorrhizal spore inoculum when replanting forest lands"[2] For this process it is best to match native trees with native mycorrhizal fungi. This method keeps and will promote the functioning of the native ecosystem, and native biodiversity.

It is assumed in a functioning forest ecosystem an underground mycelial network persists even if no fruiting bodies are visible.[4] A period of disappearance of mushrooms from an area should not cause alarm. In order to trigger the formation of fruiting bodies, many fungal species require specific environmental conditions. Most species of fungi do not fruit year round.

Mycoforestry is an emergent scientific field and practice.[2] Until broadly standardized protocols are created and perfected, the collection of both current and historical ecological site conditions will improve the success of the project.[2] Therefore, a survey of fungal relations at the site under both prime and poor conditions is beneficial to implementation of a mycoforestry system.

16.2   Saprotrophic fungi

The second principle is to promote saprotrophic fungi in the environment.[2] Saprophytic fungi are crucial to mycoforestry systems because these are the primary composers breaking down wood and returning nutrients to the soil for use by the rest of the forest ecosystem. This can be accomplished through inoculation of wood debris at site. Spored oils can be used in chainsaws when problematic or invasive hardwood requires felling. This method is a simple means to inoculate a tree. Additionally plug spawn can be implemented and injected into wood mass again prompting col-

Edible oyster mushrooms (Pleurotus sp.) fruiting from a stump

onization by the selected fungus. Eventually repeated colonization efforts should not be necessary as many fungal life forms are strong and will spread and sustain in the soil on their own.[4]

In management of the mycoforestry system it is important that dead wood be in contact with the ground. This allows fungus to reach up from the soil and decompose fallen wood releasing nutrients at a much quicker rate then if the wood is left standing.[2] Additionally it is important to leave dead wood on site for decomposition back into the soil.[2] This philosophy is similarly based to the fact that clear cutting of a forest reduces soil nutrients and thickness.[3]

16.3 Beneficial fungal interactions

Armillaria, *a parasitic fungus*

The third principal is to implement species known to benefit plant species.[2] These are commonly mycorrhizal fungus that form long term associations with plants, often extend-

ing inside of plants roots acting as an additional root system providing for better absorption of nutrients and water.

Utilizing mushroom species that attract insects could be a useful source of fish food. This practice makes the mycoforestry a larger system. Unlike most agriculture systems it helps the environment in a number of ways. It ties all biological aspects of the environment together, creating sustainable living and food production as well as sustainable fisheries similar to the ancient Hawaiian Ahupua'a, which utilized sustainable all portions of the land for environmental and food security.

Additionally fungal species can be implemented that compete with disease causing agents like Armillaria root rots[2] to provide long term protection of the forestry system.

Additionally, the implementation of an agroforestry system performs mycoremediation and mycofiltration activities cleaning up toxins and restoring the environment.

16.4 See also

- Forestry portal

16.5 References

[1] Friedman, Zev. Digging In. *New Life Journal.* 1 May 2009.

[2] Stamets, Paul (2005). *Mycelium running: how mushrooms can help save the world*. Ten Speed Press. ISBN 1-58008-579-2.

[3] Dahlgren, R. A.; Driscoll, C. T. The effects of whole-tree clear-cutting on soil processes at the Hubbard Brook Experimental Forest, New Hampshire, USA. *Plant and Soil.* Volume 158, Number 2 / January 1994.

[4] Frankland, Juliet C. All you ever wanted to know about Mycelium. *NWFG Newsletter.* April 1997. (ISSN 1465-8054) Print.

16.6 External links

- Spinosa, Ron. Fungi and Sustainability. *Fungi magazine.* Spring 2008.

- Stamets, Paul. Mycotechnology. *Fungi Perfecti.*

Chapter 17

Permaforestry

Permaforestry is an approach to the wildcrafting and harvesting of the forest biomass that uses cultivation to improve the natural harmonious systems. It is a relationship of interdependence between humans and the natural systems in which the amount of biomass available from the forest increases with the health of its natural systems.

Examples of bioproducts derived from biomass created through permaforestry include: honey, maple syrup and other tree saps, gourmet foods, functional foods, berries, wild mushrooms, acorns, walnuts, mockernut, pignut hickory, ginseng, wild rice, herbs, fiddleheads, extracting pinenuts from pinecones, fish, frogs and crustaceans, deer, moose, brought over wild boar (to the Americas), partridge, grouse, free living turkey, Canada geese and co. (other geese), rabbit, natural furs, partridge berry, bearberry, snowberry, lindenberry, mulberries, juneberries, gooseberries, (and more), leeks, spring beauties, wood sorrel, wood strawberry, burdock, dandelions, pharmaceuticals, natural health products, essential oils, educational products, arts and crafts, decorative products, floral and greenery, garden horticultural products, woodworking, lumber, biochemicals, candle wax from candleberry resin which you melt off, basswood tree leaves, greenbriar, and hydrofracking for carbon clusters and hydrocarbon chains (through use of water jets).

17.1 History

Permaforestry was extensively practiced by many aboriginal cultures throughout the world prior to colonization. It was replaced by industrial agriculture in most regions where the land could permit the use of machinery, monoculture, or intensive farming and harvesting practices. In the beginning of the 21st century there was a new surge of interest in permaforestry practices to address social issues such as food shortages, rural impoverishment, and changes in the logging industry. Furthermore, climate change and the "green" shift have inspired many individuals to revisit the old resource production methods that worked with nature rather than against it. The high price of agricultural land and machinery has also contributed to the development of permaforestry on land that had been previously classified as unsuitable for agriculture.

17.2 See also

- Biomass
- Biomass (ecology)
- Bioproduct
- Native
- Natural landscape
- Permaculture
- Terra preta
- Traditional ecological knowledge
- Wildness
- Wilderness
- Wildlife
- World Forestry Congress

Chapter 18

Polyculture

Polyculture providing useful within-field diversity: companion planting of carrots and onions. The onion smell puts off carrot root fly, while the smell of carrots puts off onion fly.[1]

Polyculture is agriculture using multiple crops in the same space, providing crop diversity in imitation of the diversity of natural ecosystems, and avoiding large stands of single crops, or monoculture. It includes multi-cropping, intercropping, companion planting, beneficial weeds, and alley cropping. It is the raising at the same time and place of more than one species of plant or animal. Polyculture is one of the principles of permaculture.

18.1 Advantages

Polyculture, though it often requires more labor, has two main advantages over monoculture.

Polyculture reduces susceptibility to disease. For example, a study in China showed that planting several varieties of rice in the same field increased yields by 89%, largely because of a dramatic (94%) decrease in the incidence of disease, which made pesticides redundant.[2]

Polyculture increases local biodiversity. This is one example of reconciliation ecology, or accommodating biodiver-

sity within human landscapes. This may also form part of a biological pest control program.

18.2 See also

- Agroecology
- Aquaponics
- Beneficial weeds
- Companion planting
- Forest gardening
- Heirloom plant
- Holistic management
- Home gardens
- Integrated Multi-trophic Aquaculture
- Monoculture
- Nurse crop

18.3 References

[1] "Companion Planting Guide". Thompson & Morgan. Retrieved 14 June 2016.

[2] Youyong Zhu; et al. (2000). "Genetic Diversity and Disease Control in Rice". *Nature* (406): 718–722. doi:10.1038/35021046.

18.4 External links

- Crop rotation and polyculture
- Polycultures in the Brazilian drylands

- Polyculture and disease prevention

- PolyCultures: Food Where We Live

- Integrated Polyculture Farming System

Chapter 19

Raised-bed gardening

Raised garden bed of lettuce, tomatoes, basil, marigolds, zinnias, garlic chives, zucchini.

Raised bed gardening

Raised-bed gardening is a form of gardening in which the soil is formed in three-to-four-foot-wide (1.0–1.2 m) beds, which can be of any length or shape. The soil is raised above the surrounding soil[1] (approximately six inches to waist-high), is sometimes enclosed by a frame generally made of wood, rock, or concrete blocks, and may be enriched with compost.[2] The vegetable plants are spaced in

Picardo Farm, Wedgwood neighborhood, Seattle, Washington: A community allotment garden with raised beds for the physically disabled.

geometric patterns, much closer together than in conventional row gardening.[2] The spacing is such that when the vegetables are fully grown, their leaves just barely touch each other, creating a microclimate in which weed growth is suppressed[2] and moisture is conserved.[3] Raised beds produce a variety of benefits: they extend the planting season,[2] they can reduce weeds if designed and planted properly,[2] and they reduce the need to use poor native soil. Since the gardener does not walk on the raised beds, the soil is not compacted and the roots have an easier time growing.[4] The close plant spacing and the use of compost generally result in higher yields with raised beds in comparison to conventional row gardening. Waist-high raised beds enable the elderly and physically disabled to grow vegetables without having to bend over to tend them.[4]

19.1 Overview

Raised beds lend themselves to the development of complex agriculture systems that utilize many of the principles

57

and methods of permaculture. They can be used effectively to control erosion and recycle and conserve water and nutrients by building them along contour lines on slopes.[3] This also makes more space available for intensive crop production.[2] They can be created over large areas with the use of several commonly available tractor-drawn implements and efficiently maintained, planted and harvested using hand tools.

This form of gardening is compatible with square foot gardening and companion planting.

Circular raised beds with a path to the center (a slice of the circle cut out) are called keyhole gardens.[5] Often the center has a chimney of sorts built with sticks and then lined with feedbags or grasses that allows water placed at the center to flow out into the soil and reach the plants' roots.

19.1.1   Materials and construction

Vegetable garden bed construction materials should be chosen carefully. Some concerns exist regarding the use of pressure-treated timber.[6] Pine that was treated using chromated copper arsenate or CCA, a toxic chemical mix for preserving timber that may leach chemicals into the soil which in turn can be drawn up into the plants, is a concern for vegetable growers, where part or all of the plant is eaten. If using timber to raise the garden bed, ensure that it is an untreated[2] hardwood to prevent the risk of chemicals leaching into the soil. A common approach is to use timber sleepers joined with steel rods to hold them together. Another approach is to use concrete blocks, although less aesthetically pleasing, they are inexpensive to source and easy to use. On the market are also prefab raised garden bed solutions which are made from long lasting polyethylene that is UV stabilized and food grade so it will not leach undesirable chemicals into the soil or deteriorate in the elements. A double skinned wall provides an air pocket of insulation that minimizes the temperature fluctuations and drying out of the soil in the garden bed. Sometimes raised bed gardens are covered with clear plastic to protect the crops from wind and strong rains.[4] Pre-manufactured raised bed gardening boxes also exist.[1] There are variants of wood, metal, stone and plastic. Each material type has advantages and disadvantages.[7]

19.2   See also

- Kitchen garden

- Therapeutic garden

- Hügelkultur: another type of raised bed

Raised garden beds with painted wooden edgings at Wise Words Community Garden in Mid-City, New Orleans

- Waru Waru – A traditional Quechua, pre-Inca system involving raised beds

- Square foot gardening

19.3   References

[1] Hughes, Megan McConnell (2010). *Better Homes & Gardens Vegetable, Fruit & Herb Gardening*. Wiley. pp. 68–69. Retrieved March 2, 2012. ISBN 978-0-470-63856-9

[2] Nones, Raymond (2010). *Raised-Bed Vegetable Gardening Made Simple*. Countryman Press. Retrieved March 2, 2012. ISBN 978-0-88150-896-3

[3] Millarville Horticultural Club (1982). *Gardening under the arch: homespun hints and money saving tips from the rigorous high country of Alberta's chinook zone*. The Club. pp. 291–292. Retrieved March 2, 2012. ISBN 0-88925-406-0

[4] Whiting, David E. (1991). *The desert shall blossom: a comprehensive guide to vegetable gardening in the Mountain West*. Horizon. pp. 41–42. Retrieved March 2, 2012. ISBN 0-88290-418-3

[5] Kemery, Ricky (January 29, 2012). "Unlock your creativity with keyhole garden". *The Journal Gazette* (Fort Wayne, IN). Retrieved March 6, 2012. External link in |publisher= (help)

[6] Lively, Ruth. "Does Pressure-Treated Wood Belong in Your Garden?". Fine Gardening Magazine. Retrieved March 6, 2012. External link in |publisher= (help)

[7] "Hochbeet kaufen - Hochbeete im Test und Vergleich". *Hochbeete kaufen* (in German). Retrieved 2016-10-09.

19.4 Bibliography

- Bird, Christopher (2001). *Cubed Foot Gardening: Growing Vegetables in Raised, Intensive Beds*. Lyons Press. Retrieved March 2, 2012. ISBN 1-58574-312-7

- Linhart, Rita & Richardson, Antoinette (2012). "Raised Bed Gardening - low cost, high yield and simply done". Books on Demand. Retrieved March 14, 2012. ISBN 978-3-8370-1841-7

19.5 External links

- The Synergistic Garden—A video by Emilia Hazelip, which provides practical information on how to garden with raised beds.

- Appeal: Keyhole gardening saves lives in world's most eroded land

Chapter 20

Regenerative agriculture

Biodiversity

Hoverfly at work

Regenerative agriculture is a sub-sector practice of organic farming designed to build soil health or to regenerate unhealthy soils. The practices associated with regenerative agriculture are those identified with other approaches to organic farming, including maintaining a high percentage of organic matter in soils, minimum tillage, biodiversity, composting, mulching, crop rotation, cover crops, and green manures.

20.1 V. organic agriculture

In the past, regenerative farming was seen as a long-term integrated approach that proponents used to build soil health, promote nutrient retention, and encourage pest and disease resistance. Many of the practices associated with regenerative farming are management practices associated with organic agriculture. In practice, these practices can be applied in any type of horticulture and properly managed livestock with Holistic Planned Grazing (Savory & Butterfield Holistic Management - A new decision making framework) where one of the main goals is to build soil organic matter, an organic practice understood by practitioners of organic farming to have far reaching benefits for plant health and

farm sustainability. When combined with the spirit of organic agriculture such practices are said to produce healthy soil, healthy food, clean water and clean air using inexpensive inputs local to the farm. Practices that minimize biota disturbance and erosion losses while incorporating carbon rich amendments and retaining the biomass of roots and shoots are encouraged in regenerative farming.[1]

20.1.1 Best practices

Foremost among best practices in regenerative farming are zero-tolerance for synthetic pesticides, fertilizers, and other inputs that disrupt soil life. On the other hand, conservation tillage, while not yet widely used in organic systems, is viewed as a regenerative organic practice integral to soil-carbon sequestration.[2]

Rodale Institute, Test Garden

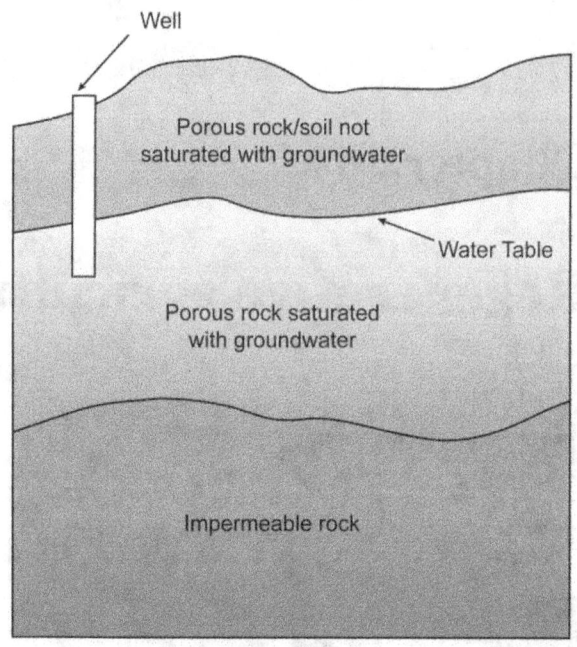

Groundwater

20.2 Contributions from The Rodale Institute

Recent scientific research has shown that practices inherent in the regenerative philosophy contribute to carbon sequestration by natural processes of photosynthetic removal and retention of atmospheric CO_2 in soil organic matter. At the forefront of this research is the Rodale Institute, which is one of the major proponents of regenerative agriculture. Notably, the concept of *regenerative organic agriculture* was coined by Robert Rodale prior to his untimely death in 1990. The Rodale approach defines regenerative farming as a long-term, holistic design that attempts to grow as much food using as few resources as possible in a way that revitalizes the soil rather than depleting it, while offering a solution to carbon sequestration. The mantra is the slogan: "Healthy Soil = Healthy Food = Healthy People."[3] In Rodale's view, when coupled with the management goal of carbon sequestration, regenerative "farming becomes, once again, a knowledge intensive enterprise, rather than a chemical and capital-intensive one."[4]

20.3 Permaculture and regeneration science

In permaculture, a regenerative farm is one where biological production and ecological structure are growing increasingly more complex over time, but yields continue to increase while external inputs decrease.[5] The way that regeneration is determined in this construct is by whether the components in the system or actions taken in the system increase both biological diversity *and* biomass. The overall health of the ecological system in which the farm and its humans are guests is determined by the health of the water and soil. This is achieved by strengthening (making more resilient, redundant) three key components of regenerative ecosystem management:

1. Increased biomass

2. Increased biological activity

3. Intentional remineralization

20.4 Foundations

The foundation of regenerative agriculture has occurred over thousands of years, a span of time when most farming was organically practiced worldwide. Even then, its knowledge base was not global because its practitioners excluded traditional non-western local knowledge systems, about which we are only beginning to learn more. It is not possible to list the many contributors or contributions to this process over many centuries. Yet, some voices and systems should be illuminated. Among these, in alphabetical order, are the following:

William Albrecht (1888–1974), an agronomist at the University of Missouri for many years. Through his writings, lectures, and radio programs, Albrecht promoted the intimate relationship between healthy soil and animal nutrition, a relationship that includes humans. Feed the soil to feed the plants to feed the consumer was his mantra.[6]

Lady Evelyn Barbara "Eve" Balfour (1899–1990) was a founding figure in the 20th century organic movement and an organic farming pioneer. She was one of the first women to study agriculture at an English university, graduating from the University of Reading, and began farming in 1920. By 1939, she had launched a privately funded experimental farm, called Haughley, to test the principles of organic farming. The initial findings of the work there and her research were published in Living Soil (1948), which has become an organic farming classic. Haughley was the first long-term comparative research project measuring results from organic and chemically based farming. What the results of her work at Haughley revealed should have led to a global push to institute sustainable agriculture throughout the British empire and beyond, but that was not to be, as the global power of post-WWII capitalism undermined such efforts.[7]

George Washington Carver (1864–1943), inventor and scientist of Tuskegee University, was the grand-daddy of sustainable agriculture in the USA. His contributions are too many to detail here, but among them is his development of the U.S. agriculture extension system and his many inventions in polymer science. His influence on U.S. agriculture in the first half of the 20th century is not fully appreciated nor, more tragically, even known by contemporary proponents of today's biointensive farming methods. Carver's scientific research was the foundation of US agriculture policy under two generations of Secretaries of Agriculture. His death in 1948 marked the end of organic agricultural practices in commercial agriculture in the United States until recently, as petrochemical companies and Agribusiness corporations competed successfully to control the actors in the various industries.

George Chan (born 1923), former Ministry of Agro-Industry & Fisheries of Mauritius, and Mae-Wan Ho, Director of the Institute of Science in Society, are architects of Dream Farm, a model of a sustainable zero-emission, zero-waste production farm that maximizes the use of renewable energies, turns residues into food and energy resources, and eliminates the need for fossil fuels. A core principle of Dream Farm is that sustainable systems are organisms, comprising the farmer, livestock and crops. In a Dream Farm, very little input is wasted or exported to the environment; most are recycled and kept inside the system by consciously integrating food and energy production. Mr. Chan has been a pioneer in the field of sustainable recycling of waste using agro-ecological systems. He developed the Integrated Farming and Waste Management System, which involves a sustainable cycle in which matter and energy circulate through different stages, dramatically increasing yields.[8]

Masanobu Fukuoka (1913–2008) of Japan lived a long and dutiful life in partnership with his environment. He was a farmer, activist, and author of the practices and theory of natural farming, on which he based his four core uncompromising principles: no cultivation, no (chemical) fer-

Field Hamois Belgium Luc Viatour

tilizers, no weeding, and no pesticides. Among his teachings is that for a farmer to be successful, he must form a partnership with the natural environment, derive an intimate understanding of it together with the plants a farmer chooses to grow. Features of his philosophy are present in most contemporary farming practices. His "seed balls" cultivation innovation is widely used in many horticultural environments and in commercial retail products, including lawn seed.[9]

Takao Furuno (born 1950) of Japan is the architect of the Aigamo Method, a modernization of an 800-year-old Chinese technique of using ducks to promote sustainable rice cultivation. Funuro's system is polycultural, combining rice cultivation with duck husbandry, aquaculture, and vegetable production. Together, these enterprises provide income from rice, vegetable and flower production, eggs, meat, and live ducks from the aigamo, and fish from the paddies. The aigamo ducks are a breed derived from wild and domestic ducks, whose ducklings provide the labor for cultivation, pest control, and manure for fertilization of the rice paddies. This beneficial marriage between duck and rice eliminates dependency on chemical fertilizers, pesticides, molluscicides, fossil fuels, and heavy duty equipment as costs, while sustaining a safe environment for farmers to work and simultaneously increasing net production and farm income.[10]

John D. Hamaker (1914–1994) was a mechanical engineer, ecologist and practical visionary who from 1968 to 1994 tried to awaken the world to the crucial need to remineralize the soils and regenerate the Earth's life-support system. His motivations included a profound desire to help create a healthy, just and real civilization rooted in ecological wisdom, and his realization that malnutrition and disease followed by famine and glaciation could be ended by a total human commitment.[11]

Julius Hensel (1844–1903), a German miller and author of Bread from Stones. In the 1890s, Hensel was an early advocate of restoring trace minerals to soil with dust from primeval stones and reported successful results with his steinmehl (stonemeal). It is said that his ideas were not accepted due to both technical limitations and financing. But, according to proponents of his method, Hensel's opposition from manufacturers of chemical fertilizers, set the stage for what would happen eventually in American agriculture following WWII when there was little opposition and the rise of the U.S. petrochemical industry. Today, Hensel's pioneer work in opposing the use of chemicals in agriculture found rebirth in the Organic Movement a half a century later. Yet, Hensel may be more modern than the most modern agricultural reformer. On the basis of theoretical chemical considerations, supported by practical tests, he claimed that his rock dust can replace not only chemical fertilizers, but all animal ones as well.[12]

Sepp Holzer (born 1942) of Austria, and his wife Veronika, have created a diverse and natural way of growing food in an unconventional way by using a terraced system of mounds on the Austrian mountainsides, referred to as hugelkultur. The mounds are built on a foundation of organic materials, a traditional way of growing in the region of The Krameterhof in Lungau, Austria, just not at 1000+ meters above sea level. Holzer's edible microclimates are considered one of the few perfectly working permaculture systems in the world. After almost 40 years of continuous production, the Holzer farms contain a complex of pond culture, terraces, water power station, thousands of fruit trees cultivated among companionable plant families, thirty different types of potatoes, many different grains, fruits, vegetables, herbs and wildflowers are growing just about everywhere — in the forest, on extremely steep hills, on rocky outcrops, on stone pathways, and around ponds, all without the use of any pesticides, herbicides or synthetic fertilizers.[13] Their video, *Farming With Nature: A Case Study of Successful Temperate Permaculture*, is widely distributed on the Internet and a must-view for anyone looking to develop a sustainable collaborative system of food production.[14]

Sir Albert Howard (1873–1947), an English botanist, was an agricultural advisor to the British government in charge of a colonial research farm at Indore in India. Howard has been called the father of modern composting for his refinement of a traditional Indian composting system into what is now known as the Indore method. It was at Indore that Howard documented and tested Indian organic farming techniques. Sir Howard shared this knowledge through the Soil Association in England and the Rodale Institute in the United States. In his later years, he was the editor of the influential journal, *Biodynamics Journal*.[15]

Elaine Ingham, a microbiologist and founder of Soil Foodweb, Inc. in 1996, is recognized as a premier authority in soil microbiology and the soil food web. Through her pioneering research and lectures, Dr. Ingham has been instrumental in popularizing the importance of soil health and the growing public understanding of the soil food web in sustaining this health. Since January 2011, she has been the Chief Scientist at The Rodale Institute where she continues to study the microbial life of the soil and to give lectures on her findings.[16]

John Ikerd, a successor to William Albrecht and professor emeritus at the University of Missouri, continues to be a staunch advocate for the "small" family farm and farmers. Today, he is an active crusader for sustainability in the US food system. His views and writings are available on his Website and at YouTube.[17] Ikerd is author of *The essentials of economic sustainability*,[18] *Small Farms are Real Farms: Sustaining People through Agriculture*[19] and *Sustainable Capitalism* (2005).[20]

John Jeavons, the inspiration and architect of a sustainable 8-step food production method officially known as Grow Biointensive, which combines elements of French intensive and biodynamics techniques. The Grow Biointensive approach is promoted by Ecology Action, a non-profit that operates a research mini-farm in Willits, CA and a retail store called Bountiful Gardens. Both projects promote the Grow Biointensive method teaching people in more than a hundred countries. Ecology Action's research and publications have several goals: (1) enabling small-scale farms and farmers worldwide to significantly increase food production and income by (2) utilizing predominantly local, renewable resources to decrease expenses and energy inputs (labor, land, water) and (3) building fertile topsoil at a rate 60 times faster than in nature.

Patricia Lanza was born in 1935 in Crossville, Tennessee, to teenagers George and Mamie Neal. An only child for eight years, she spent her early formative years with grandparents while her parents worked in Detroit, MI. She perfected and authored several books about lasagna gardening or sheet mulch gardening, a type of gardening perfected by Ruth Stout in the 1950s, but whose books had gone out of print.[21] Lanza introduced a new generation of gardeners to the ultimate no-till method of growing in which little or no labor is wasted digging and amending soils. Instead, the lasagna method feeds the soil biota from above and encourages the soil food web to do the work of aerating and mixing the nutrients into the soil below.[22]

Jacob Mittleider, architect of the Mittleider Method, a popular contemporary method of soil-less growing. Mittleider's system is being continued and refined by Jim Kennard at the Food for Everyone Foundation.[23]

Bill Mollison and **David Holmgren**, architects of permaculture or "permanent agriculture", a holistic approach that combines ecological design with natural principles of horticultural production. Both men are ardent advocates for creating communities that work in harmony with nature rather than in opposition to it. Permaculture is a worldwide phenomenon whose principles and practices have been published in many of the world's languages and is rapidly being integrated into public planning projects across the globe. Video presentations about Permaculture and its practitioners are broadly distributed on the Internet and can be found at YouTube.[24]

Maynard Murray (1910–1983) was a medical doctor and a pioneer in merging the disciplines of biology, health and agriculture from the 1930s when he began experimenting with "sea-solids"–mineral salts that remain after total sea water evaporation. Around 1940, he began to perform extensive experiments to determine when the proportions of trace minerals and other elements present in sea water were optimum for growth and health of both land and sea life. His extensive experiments demonstrated repeatedly and conclusively that plants fertilized with sea solids and animals fed sea-solid-fertilized feeds grow stronger and more resistant to disease. Murray recounts his experiments and presents his conclusions in his classic work, Sea Energy Agriculture (1976; republished in 2001). Largely ignored during his lifetime, his lifelong quest contributed greatly to our understanding of the role of trace minerals in the healthy growth of all organisms on the planet, including humans.[25]

J.I. Rodale (1898–1971) was an early proponent of organic and regenerative farming and founder of the Rodale Institute in the United States. He is credited with launching organic gardening practices more broadly in the United States through his writings, research, and publishing enterprise. The Rodale enterprises continue to make contributions around the world through their advocacy, research, demonstrations, and publishing.

Robert Rodale (1930–1990), former CEO of The Rodale Institute, was a major advocate of regenerative agriculture, fostered the Regenerative Agriculture Association, published numerous books on the subject, funded research, established demonstration fields, sponsored practitioners in the field, and spread the knowledge system of regenerative agriculture around the globe. He coined the concept of 'regenerative organic agriculture' to distinguish it from 'sus-

tainable' agriculture.

Bhaskar H. Save (born 1922) is creator of the highly successful Kalpavruksha ("wish-fulfilling tree") Farm in Umbergaon, India established in 1953. After practicing traditional agriculture for many years with poor results, Sri Save committed his resources to organic farming and developed a system of natural farming that Masanobu Fukuoka, the noted founder of natural farming, praised as the best example of natural farming he had witnessed anywhere. Sri Save used intensive interplanting in which short life-span vegetables (alpa-jeevi), medium life-span species (madhya-jeevi – such as banana, papaya, and custard apple), and long life-span species (deergha-jeevi–such as chikoo, coconut, mango) are combined and phased in over time until the long life-span species mature.

Ethan Roland Soloviev and **Gregory Landua**, co-founders of Terra Genesis International (a regenerative agriculture and supply company), published a paper in 2016 titled *Levels of Regenerative Agriculture.*[26][27] In this paper, they describe a four-fold framework consisting of:

- Functional Regenerative Agriculture: "humans can do good through their agricultural production"

- Integrative Regenerative Agriculture: "grow the health and vitality of the whole ecosystem"

- Systemic Regenerative Agriculture: requiring personal development; "farms are woven into an ecosystem of enterprises operating in their bioregion"

- Evolutionary Regenerative Agriculture: requiring pattern understanding; "harmonize with the potential of a place," and "develop a diversity of global and local regenerative producer webs"

Rather than creating a hierarchy, Soloviev and Landua posit that each level of regenerative agriculture has its place, depending upon context and aim.

Ruth Stout (1884–1980) lived a long, active and productive life. By the 1950s, she had perfected a "no-till" method of gardening that she promoted as "no work" in her writings about gardening, including two books, *How to Have a Green Thumb Without an Aching Back* and *Gardening Without Work for the Aging, the Busy, and the Indolent*. The latter volume was republished by Mother Earth News in 2011.[28] Her work has led to other innovations in no-till practices, such as *slash and mulch* in the tropics.

Charles Walters (1926–2009) was an economist, journalist, farmers advocate in the first phase of his career with the National Farmer's Organization; and founder, publisher and editor of Acres U.S.A., North America's oldest publisher on production-scale organic and sustainable farming in the second phase of his extraordinary life. Walters penned hundreds of articles on the technologies of organic and sustainable agriculture and is author or co-author of several books, including *Eco-Farm*, *Weeds: Control Without Poisons*, *Unforgiven*, a book about visionary farm economist Carl Wilken, and many more. In 1970, shortly after he started Acres, Walters coined the term "eco-agriculture" because he wanted to unify the concepts of "ecological" and "economical" in the belief that unless agriculture was ecological, it could not be economical.[29]

Keyline Irrigation, Taranaki Farm

Don Weaver, a protégé of John Hamaker, is an ecologist and gardener, who assisted Hamaker in advocating for policies and practices of soil remineralization, biosphere regeneration, and climatic stabilization. He continues to promote these causes today.[30]

Booker T. Whatley (1915–2005), a horticulturalist and beneficiary of the George Washington Carver tradition, may be best remembered for popularizing U-Pick farms and their direct marketing approach through fee-based customer subscriptions. But, he was also among the first practitioners of sustainable agriculture to focus more directly on the economic concerns of small farmers, encouraging them to identify high value crops and enterprises that were more profitable on smaller units of land, such as shiitake mushrooms, the husbandry of small ruminants, specialty cheeses, and much more.[31]

P.A. Yeomans (1904–1984), an Australian geologist and the architect of the Keyline design, an innovative solution to farm water management, is little known outside of Australia for his many contributions from sustainable agriculture to soil fertility to farm management in the early 20th century. His use of land topography to harvest rain water into ponds is quietly used today in the construction of swales and berms in production units ranging from backyard gardens to the monumental landscaping practices of Sepp Holzer today.[32]

20.5 See also

- Agroecological restoration
- Agroecology
- Agroforestry
- Biointensive agriculture
- Farmer-managed natural regeneration
- Permaculture
- Zaï

20.6 References

[1] "Regenerative Organic Agriculture and Climate Change: A Down-to-Earth Solution to Global Warming." Rodale Institute, 2014. p. 8.

[2] "Regenerative Organic Agriculture and Climate Change: A Down-to-Earth Solution to Global Warming." Rodale Institute, 2014. pp.7-10.

[3] "Regenerative Farming." PowerPoint presentation, The Rodale Institute, 12 Feb 2003.

[4] "Regenerative Organic Agriculture and Climate Change: A Down-to-Earth Solution to Global Warming." Rodale Institute, 2014. pp. 7-8.

[5] Ben Falk, *The resilient farm and homestead*. Chelsea Green, 2013. p. 280.

[6] Biographical Profile of William Albrecht

[7] Balfour, Lady Eve. *9,600 Miles Through The U.S.A. in a Station Wagon*. London: The Soil Association, 1954.

[8] Dream Farms ISIS Report 09/06/05; How to Beat Climate Change & Be Food and Energy Rich - Dream Farm 2". ISIS Report 10/07/07.

[9] Fukuoka, Masanobu et. al. *The One-Straw Revolution: An Introduction to Natural Farming* New York Review Books, 2009 and *Natural Way of Farming: The Theory and Practice of Green Philosophy*. Other India Press, 1985. 284p.

[10] Furuno, Takao. *The Power of Duck: Integrated Rice and Duck Farming*. Tagari Publications, 2002.

[11] Hamaker, John D. and Donald Weaver. *The Survival of Civilization*. Hamaker-Weaver Publishers, 1982. 234p. Reprinted, 2002.

[12] Hensel, Julius. *Bread From Stones: A New and Rational System of Land Fertilization and Physical Regeneration*. Republished by Acres USA, Austin, Texas, 1991. 102p.

[13] Holzer, Sepp. Sepp Holzer's Permaculture: A practical guide to small-scale, integrative farming and gardening. White River Junction, VT: Chelsea Green Publishing, 2011. xix, 246p.

[14] Sepp Holzer and Permaculture videos

[15] Howard, Sir Albert. An Agricultural Testament. London: Oxford University Press, 1943.

[16] Ingram, Elaine. (2000) *Soil Biology Primer*. USDA.

[17] John E. Ikerd Website; [http://web.missouri.edu/ikerdj/ University of Missouri-Faculty Emeritus

[18] The essentials of economic sustainability Lynne Rienner Publishers, 2012.

[19] Small Farms are Real Farms. Acres, 2007

[20] Sustainable Capitalism: A Matter of Common Sense. Lynne Rienner Publishers, 2005.

[21] Mother Earth News. (article)

[22] Lanza, Patricia. *Lasagna Gardening: A New Layering System for Bountiful Gardens: No Digging, No Tilling,No Weeding, No Kidding!* Rodale Books, 1999. 256p.

[23] Mittleider, Jacob. *Food For Everyone: The Mittleider Method*. Color Press. n.d, 624p.; *Mittleider Grow-Box Gardens*. International Food Production Methods, Inc., 1975. 195p.

[24] Mollison, Bill. *Permaculture: A Designer's Manual*. Tagari Publications, 1988. 576p; Holmgren, David. Permaculture: Principles and Pathways beyond Sustainability. *Holmgren Design Services, 2002. 320p.* .

[25] Murray, Maynard. *Sea energy agriculture*. 2nd ed. revised. Austin, TX: Acres, USA, 2003. vii, 109p. Nauta, Phil. Building soils naturally. Austin, TX: Acres, USA, 2012. xvi, 303p.

[26] "White Paper: Levels of Regenerative Agriculture". *www. terra-genesis.com*. Retrieved 2016-10-13.

[27] Roland Soloviev, Ethan (20 September 2016). "What is Regenerative Agriculture?". Quora. Retrieved 13 October 2016.

[28] Stout, Ruth. Gardening without Work.

[29] "What is eco-agriculture."

[30] Hamaker, John D. and Donald Weaver. *The Survival of Civilization*. Hamaker-Weaver Publishers, 1982. 234p. Reprinted, 2002.

[31] Whatley, Booker T. *How to Make $100,000 Farming 25 Acres*. Emmaus, Pennsylvania, Regenerative Agriculture Association, 1987. 180 pages.

[32] Yeomans, P.A. *The keyline plan*. Sydney: P.A. Yeomans, 1954. [Source: The Holistic Agriculture Library and *The challenge of landscape*. Sydney: Keyline Publishing PTY, Ltd., 1958. [Source: The Holistic Agriculture Library]

Chapter 21

Sheet mulching

This article is about mulching in permaculture. For the similar but more general technique, see Mulch.

In permaculture, **sheet mulching** is an agricultural no-dig gardening technique that attempts to mimic natural forests' processes. When deployed properly and in combination with other permacultural principles, it can generate healthy, productive, and low maintenance ecosystems.[1][2]

21.1 Technique

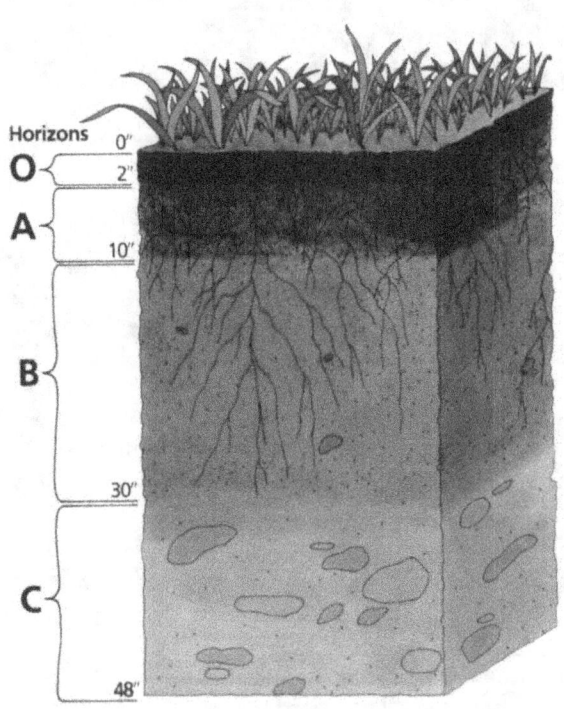

Typical layers of natural soil.

A model for sheet mulching consists of the following steps:[1][3]

1. The area of interest is flattened by trimming down existing plant species such as grasses.

2. The soil is analyzed and its pH is adjusted (if needed).

3. The soil is moisturized (if needed) to facilitate the activity of decomposers.

4. The soil is then covered with a thin layer of slowly decomposing material (known as the **weed barrier**), typically cardboard. This suppresses the weeds by blocking sunlight, adds nutrients to the soil as weed matter quickly decays beneath the barrier, and increases the mechanical stability of the growing medium.

5. A layer (around 10 cm thick) of weed-free soil, rich in nutrients is added, in an attempt to mimic the A horizon.

6. A layer (at most 15 cm thick) of weed-free, woody and leafy matter is added in an attempt to mimic the forest floor or O Horizon. Theoretically, the soil is now ready to receive the desirable plant seeds.[4]

21.2 Variations and considerations

- Often the barrier is applied a few months before planting to ensure the penetration of roots of newly planted seeds.[3]

- Very thick barriers can cause anaerobic conditions.

- Some permaculturists incorporate composting in steps 5 and/or 6.[3]

- Sheets of newspaper and clothing can be used instead of cardboard.[3]

- Before step 4, an initial layer (2–3 kg/m^2) of matter rich in nutrients (such as compost or manure) may be added to bolster decomposition.[1]

- Some varieties of grasses and weeds may be beneficial in a number of ways. Such plants can be controlled and used rather than eradicated.[1] See also: mulch, list of beneficial weeds.

- One variation of mulching, called hugelkultur, involves using buried logs and branches as the first layer of the bed.[5]

21.3   Advantages

Sheet mulch has important advantages relative to conventional methods:

- Improvement of desirable plants' health and productivity.[1]

- Retention of water and nutrients and stabilization of biochemical cycles.[1]

- Improvement of soil structure, soil life, and prevention of soil erosion.[1][6]

- Avoidance of potentially dangerous pesticides, especially herbicides.

- Reduction of overall maintenance labor and costs.[6]

21.4   Disadvantages

- Some weed seeds (such as those of Bermuda grass and species of bindweed) may persist under the barrier and within the soil seed bank.[3]

- Termites are attracted to the area. While they are a natural part of the ecosystem that transforms the weed barrier into rich soil, they can pose a hazard to nearby wood-framed structures.

- Slug populations may increase during the early stages of decomposition. However they can be kept away or harvested.[3]

- The system may need a constant supply of organic material, at least during the early stages.[1]

- Roaming animals may interrupt the sheet mulching process.[1]

21.5   See also

- Agroecology

- Ecoagriculture

- Ecological design

- Ecosystem approach

- Forest gardening

21.6   References

[1] Elevitch, Craig; Wilkinson, Kim (1998). "Sheet Mulching: Greater Plant and Soil Health for Less Work". Agroforestry.net. Retrieved November 9, 2011.

[2] Mason, John (2003). *Sustainable Agriculture* (Second ed.). Collingwood, Vic.: CSIRO. ISBN 978-0-643-06876-6.

[3] Hemenway, Toby (2009). *Gaia's Garden: A Guide to Home-Scale Permaculture* (2nd ed.). White River Junction, Vt.: Chelsea Green Pub. ISBN 978-1-60358-029-8.

[4] Stout, Ruth (February–March 2004). "Ruth Stout's System". *Mother Earth News*. Retrieved November 9, 2011.

[5] Nicole Faires, *The Ultimate Guide to Permaculture*, Skyhorse Publishing Inc., p. 228, ISBN 978-1-62087-316-8, retrieved 3 July 2012

[6] "Sheet Mulching". StopWaste.Org. Retrieved November 9, 2011.

Chapter 22

Spent mushroom compost

Spent mushroom compost is the residual compost waste generated by the mushroom production industry. It is readily available (bagged, at nursery suppliers), and its formulation generally consists of a combination of wheat straw, dried blood, horse manure and ground chalk, composted together. It is an excellent source of humus, although much of its nitrogen content will have been used up by the composting and growing mushrooms. It remains, however, a good source of general nutrients (0.7% N, 0.3% P, 0.3% K plus a full range of trace elements), as well as a useful soil conditioner. However, due to its chalk content, it may be alkaline, and should not be used on acid-loving plants, nor should it be applied too frequently, as it will overly raise the soil's pH levels.[1]

Mushroom compost may also contain pesticide residues, particularly organochlorides used against the fungus gnat. If the compost pile was stored outside, it may contain grubs or other insects attracted to decaying matter. Chemicals may also have been used to treat the straw, and also to sterilize the compost. Therefore, the organic gardener must be careful regarding the sourcing of mushroom compost; if in doubt, samples can be analyzed for contamination – in the UK, the Department for Environment, Food and Rural Affairs is able to advise regarding this issue.

Commercially available 'spent' mushroom compost is not always truly spent. It is sold by mushroom farms when it is no longer producing commercially viable yields of mushrooms. It can be used to grow further smaller crops of mushrooms before final use on the garden.

22.2   External links

- How to grow mushrooms in spent mushroom compost before using in garden beds

22.1   References

[1] Bradley, Steve (2004). *Vegetable Gardening: Growing and Harvesting Vegetables*. Murdoch Books. ISBN 1-74045-519-3.

Chapter 23

Synergistic gardening

Synergistic gardening is a system of organic gardening, developed by Emilia Hazelip. The system is strongly influenced by permaculture, as well as the work of Masanobu Fukuoka and Marc Bonfils.[1] After establishing the garden, there is no further digging, ploughing or tilling, and no use of external inputs such as manures and other fertilizers, or pesticides. Soil health is maintained by the selection of plants, mulching, and recycling of plant residues.[2]

23.1 References

[1] "The Synergistic Garden". *Excerpt from Permaculture Magazine*. Issue 19. Spring 1999. Archived from the original on March 16, 2006. Retrieved 14 January 2013.

[2] Emilia Hazelip's emails (x105) articulating her practises and positions. Dec. 2001 – Jan. 2003

- Introduction to Synergistic Gardening

Chapter 24

Three Sisters (agriculture)

This article is about maize, beans, and squash. For the variety of tomato, see Three Sisters tomato.

The **Three Sisters** are the three main agricultural crops of

Three Sisters as featured on the reverse of the 2009 Native American U.S. dollar coin

various Native American groups in North America: winter squash, maize (corn), and climbing beans (typically tepary beans or common beans). The Iroquois, among others, used these "Three Sisters" as trade goods.

In a technique known as companion planting, the three crops are planted close together. Flat-topped mounds of soil are built for each cluster of crops.[1] Each mound is about 30 cm (12 in) high and 50 cm (20 in) wide, and several maize seeds are planted close together in the center of each mound. In parts of the Atlantic Northeast, rotten fish or eels are buried in the mound with the maize seeds, to act as additional fertilizer where the soil is poor.[2] When the maize is 15 cm (6 inches) tall, beans and squash are planted around the maize, alternating between the two kinds of seeds. The process to develop this agricultural knowl-

edge took place over 5,000–6,500 years. Squash was domesticated first, with maize second and then beans being domesticated.[3][4] Squash was first domesticated 8,000–10,000 years ago.[5][6]

The three crops benefit from each other. The maize provides a structure for the beans to climb, eliminating the need for poles. The beans provide the nitrogen to the soil that the other plants use, and the squash spreads along the ground, blocking the sunlight, helping prevent the establishment of weeds. The squash leaves also act as a "living mulch", creating a microclimate to retain moisture in the soil, and the prickly hairs of the vine deter pests. Corn, beans, and squash contain complex carbohydrates, essential fatty acids and all eight essential amino acids, allowing most Native American tribes to thrive on a plant-based diet.[7]

Native Americans throughout North America are known for growing variations of Three Sisters gardens.[8] The milpas of Mesoamerica are farms or gardens that employ companion planting on a larger scale.[9] The Anasazi are known for adopting this garden design in a drier environment. The Tewa and other Southwestern United States tribes often included a "fourth sister" known as "Rocky Mountain bee plant" (*Cleome serrulata*), which attracts bees to help pollinate the beans and squash.[10]

The Three Sisters planting method is featured on the reverse of the 2009 US Sacagawea Native American dollar coin.[11]

24.1 Gallery

- Maize
- *Phaseolus vulgaris*
- Butternut squash, a type of winter squash.

24.2 See also

- Agriculture in the prehistoric Southwest

- Agroforestry

- Companion planting

- Eastern Agricultural Complex

- Intercropping

- Milpa

- Multiple cropping

- Polyculture

24.3   References

[1] Mt. Pleasant, Jane (2006). "38". In John E. Staller; Robert H. Tykot; Bruce F. Benz. *The science behind the Three Sisters mound system: An agronomic assessment of an indigenous agricultural system in the northeast. Histories of Maize: Multidisciplinary approaches to the prehistory, linguistics, biogeography, domestication, and evolution of maize.* Amsterdam: Academic Press. pp. 529–537. ISBN 978-1-5987-4496-5.

[2] Vivian, John (February–March 2001). "The Three Sisters". Mother Earth News. Retrieved September 18, 2013.

[3] Landon, Amanda J. (2008). "The "How" of the Three Sisters: The Origins of Agriculture in Mesoamerica and the Human Niche". *Nebraska Anthropologist.* Lincoln, NE: University of Nebraska-Lincoln: 110–124.

[4] Bushnell, G. H. S. (1976). "The Beginning and Growth of Agriculture in Mexico". *Philosophical Transactions of the Royal Society of London.* London: Royal Society of London. **275** (936): 117–120. doi:10.1098/rstb.1976.0074.

[5] Smith, Bruce D. (May 1997). "The Initial Domestication of Cucurbita pepo in the Americas 10,000 Years Ago". *Science.* Washington, DC: American Association for the Advancement of Science. doi:10.1126/science.276.5314.932.

[6] "Cucurbitaceae--Fruits for Peons, Pilgrims, and Pharaohs". University of California at Los Angeles. Retrieved September 2, 2013.

[7] McDougall, John (2002). "Misinformation on Plant Proteins". *Circulation.* Dallas, Tx: American Heart Association. **106** (20): e148—e148.

[8] Wilson, Gilbert (1917). *Agriculture of the Hidatsa Indians: An Indian Interpretation.* Gloucestershire: Dodo Press. p. 25. ISBN 978-1409942337.

[9] Mann, Charles (2005). *1491: New Revelations of the Americas Before Columbus.* New York: Vintage Books. pp. 220–221. ISBN 978-1-4000-3205-1.

[10] Hemenway, Toby (2000). *Gaia's Garden: A Guide to Home-Scale Permaculture.* White River Junction, VT: Chelsea Green Publishing. p. 149. ISBN 1-890132-52-7.

[11] "2009 Native American $1 Coin". United States Mint. Retrieved September 18, 2013.

24.4   External links

- Companion Planting-Three Sisters, Old Farmer's Almanac

- Virtual Museum of Canada, The St. Lawrence Iroquoians — virtual exhibit that includes information on Iroquoian agriculture and the Three Sisters

Chapter 25

Tree bog

A **tree bog** or **Treebog** is a type of low-tech dry toilet. It is a hole in the ground (similar to a pit latrine), with any of a wide range of species planted around it. It can be considered an example of permaculture design, as it functions as a system for converting urine and faeces to biomass, without the need to handle excreta.

25.1 Etymology

The term "Treebog" was coined Jay Abrahams of Biologic Design. *Bog* is a British English slang word for toilet, not to be confused with its other meaning of wetland.

25.2 History

The Treebog is a simple method of composting wastes. Abrahams claims that from 1995-2011, around 1500 Treebogs may have been built in Britain.[1] They have been on sites ranging from fruit farms, pick-your-own enterprises, campsites, an angling lake, festival sites, remote/low impact dwellings, holiday cottages, allotments, and church yards where there is no mains water supply.

In 2011, Abrahams claimed that the Treebog had attracted the attention of NGOs and aid workers who hope to develop its potential for shanty towns or refugee camps - anywhere that water is scarce and the population pressure on resources is high.[2]

25.3 Plant growth

A Treebog is simply a controlled compost heap whose function has been enhanced by use of moisture or nutrient-hungry trees. They use no water, purify waste as they create a biomass resource, and also contain the organic waste material, thus preventing the spread of disease.

The main requirement is that the planted species should be nutrient-hungry. It is a bonus if they can be harvested or coppiced for productive uses, e.g. willow cultivars. Apart from willow coppice, soft fruit such as black currants and sweet-smelling herbs such as mint will thrive around a Treebog. If left unmanaged, a Treebog will soon be surrounded by weed species such as nettles, but a little management and conscious planting can create a fertile and productive bog garden.

Both the solids and liquids are deposited within the Treebog base, where the solids compost and the liquids soak through the soil. The associated dense rootzone enables the nitrogen to be rapidly absorbed and metabolised by the mycorhyzal species. The faeces are contained within the Treebog base, which is well ventilated to allow aerobic decomposition to occur, the mineralised material feeding the trees around it.

25.4 Construction

A seating platform/cubicle is mounted at least one meter high. The area beneath the seating platform is enclosed by a double-layer of chicken wire; this acts as an effective child-proof barrier and allows air to circulate through the compost heap.

Sawdust, straw, woodchip, ash or other high-carbon matter is used to balance the high-nitrogen fæces. One design used Effective Micro-organism bran, which helped keep the Treebog virtually odour free.[3]

The space between the wire is stuffed with straw, which acts as a wick to help sop up excess urine, preventing the likelihood of odour problems due to incomplete biological absorption of the nitrogen from the urine. The straw-filled wire also enables the pile to be well-aerated whilst acting as a visual screen for the first year's use.

The structure is surrounded by two closely planted rows of osier or biomass willow cuttings; this living wall of willow can then be woven into a hurdle-like structure and its annual growth can be harvested.

25.5   See also

- Ecological sanitation

- Reuse of excreta

- Arborloo, another version

25.6   References

[1] "The Making of a Biologic Treebog" (PDF). *Living Woods Magazine*: 10–13. January–February 2011. Retrieved 5 July 2016.

[2] "The Making of a Biologic Treebog" (PDF). *Living Woods Magazine*: 10–13. January–February 2011. Retrieved 5 July 2016.

[3] Tim Green (18 May 2011). "A Loo with a View - Build your own Treebog". Permaculture Magazine. Retrieved 23 February 2012.

-

- "Woodland Toilets or Tree Bogs". WoodlandsTV. 15 May 2009. Retrieved 23 February 2012.

25.7   External links

- Build your own Treebog

Chapter 26

Vegan organic gardening

Vegan organic gardening and farming is the organic cultivation and production of food crops and other crops with a minimal amount of exploitation or harm to any animal.[1] Vegan gardening and stock-free farming methods use no animal products or by-products, such as bloodmeal, fish products, bone meal, feces, or other animal-origin matter, because the production of these materials is viewed as either harming animals directly, or being associated with the exploitation and consequent suffering of animals. Some of these materials are by-products of animal husbandry, created during the process of cultivating animals for the production of meat, milk, skins, furs, entertainment, labor, or companionship; the sale of by-products decreases expenses and increases profit for those engaged in animal husbandry, and therefore helps support the animal husbandry industry, an outcome most vegans find unacceptable.[2]

26.1 Types

26.1.1 Veganiculture

Vegan - Organic - Agriculture / Permaculture The Future Of Farming! All Things Related To: Organic Gardening, Farming & Food Forests Free From Animals & Animal Products

Robert Hart's forest garden in Shropshire, England.

26.1.2 Forest Gardening

Forest gardening is a fully plant-based organic food production system based on woodland ecosystems, incorporating fruit and nut trees, shrubs, herbs, vines and perennial vegetables.[3] Making use of companion planting, these can be intermixed to grow in a succession of layers, to replicate a woodland habitat. Forest gardening can be viewed as a way to recreate the Garden of Eden.[4] The three main products from a forest garden are fruit, nuts and green leafy vegetables.[5]

Robert Hart adapted forest gardening for temperate zones during the early 1960s. Robert Hart began with a conventional smallholding at Wenlock Edge in Shropshire. However, following his adoption of a raw vegan diet for health and personal reasons, Hart replaced his farm animals with plants. He created a model forest garden from a small orchard on his farm and intended naming his gardening method *ecological horticulture* or *ecocultivation*.[6] Hart later dropped these terms once he became aware that *agroforestry* and *forest gardens* were already being used to describe similar systems in other parts of the world.[7]

26.1.3 Vegan permaculture

Vegan permaculture (also known as veganic permaculture, veganiculture, or vegaculture) avoids the use of domesticated animals.[8] It is essentially the same as permaculture except for the addition of a fourth core value; "Animal Care."[9] Zalan Glen, a raw vegan, proposes that *vegaculture* should emerge from permaculture in the same way veganism split from vegetarianism in the 1940s.[9] Vegan permaculture recognizes the importance of free-living animals, not domesticated animals, to create a balanced ecosystem.[8]

26.1.4 Veganic Gardening

The veganic gardening method is a distinct system developed by Rosa Dalziell O'Brien, Kenneth Dalziel O'Brien and May E. Bruce, although the term was originally coined by Geoffrey Rudd as a contraction of *vegetable organic* in order to "denote a clear distinction between conventional chemical based systems and organic ones based on animal manures".[10] The O'Brien system's principal argument is that animal manures are harmful to soil health rather than that their use involves exploitation of and cruelty to animals.

The system employs very specific techniques including the addition of straw and other vegetable wastes to the soil in order to maintain soil fertility. Gardeners following the system use soil-covering mulches, and employ non-compacting surface cultivation techniques using any short-handled, wide-bladed, hand hoe. They kneel when surface cultivating, placing a board under their knees to spread out the pressure, and prevent soil compaction. Kenneth Dalziel O'Brien published a description of his system in *Veganic Gardening, the Alternative System for Healthier Crops*:

> The veganic method of clearing heavily infested land is to take advantage of a plant's tendencies to move its roots nearer to the soil's surface when it is deprived of light. To make use of this principle, aided by a decaying process of the top growth of weeds, etc., it is necessary to subject such growth to heat and moisture in order to speed up the decay, and this is done by applying lime, then a heavy straw cover, and then the herbal compost activator...The following are required: Sufficient new straw to cover an area to be cleared to a depth of 3 to 4 inches.[11]

The O'Brien method also advocates minimal disturbance of the soil by tilling, the use of cover crops and green manures, the creation of permanent raised beds and permanent hard-packed paths between them, the alignment of beds along a north-south axis, and planting in double rows or more so that not every row has a path on both sides. Use of animal manure is prohibited.

26.1.5 Vegan biodynamic agriculture

The German agricultural researcher Maria Thun (1922 - 2012)[12] developed vegan equivalents to the traditional, animal based biodynamic preparations. As a reaction to the BSE scandal in Europe she started researching plant based preparations, using tree barks as replacement for animal organs as sheath for the preparations.[13]

In particular in Italy, there is a movement of vegan biodynamic farming, represented by farmers such as Sebastiano Cossia Castiglioni [14] and Cristina Menicocci.[15]

There are many other methods currently used and under development.

26.2   Practices

Soil fertility is maintained by the use of green manures, cover crops, green wastes, composted vegetable matter, and minerals. Some vegan gardeners may supplement this with human urine from vegans (which provides nitrogen) and 'humanure' from vegans, produced from compost toilets.[2] Generally only waste from vegans is used because of the expert recommendation that the risks associated with using composted waste are acceptable only if the waste is from animals or humans having a largely herbivorous diet.[16]

Veganic gardeners may prepare soil for cultivation using the same method used by conventional and organic gardeners of breaking up the soil with hand tools and power tools and allowing the weeds to decompose.

26.3   See also

- Climate-friendly gardening
- Deep ecology
- Environmental vegetarianism
- Movement for Compassionate Living
- Plants for a Future
- Stock-free agriculture

26.4   Notes

[1] "Different ways to garden veganically". Veganic Agriculture Network. 7 August 2011.

[2] "Growing without cruelty - the vegan organic approach". The Vegan Society.

[3] Kip Bellairs (7 May 2011). "Forest Gardening in a Nutshell". Veganic Agriculture Network.

[4] Graham Bell (2004). *The Permaculture Garden*, p.129, "The Forest Garden...This is the original garden of Eden. It could be your garden too."

- Also see Rob Hopkins (foreword), Martin Crawford (2010). *Creating a Forest Garden: Working with Nature to Grow Edible Crops*, p.10 "Perhaps what Hart created was the closest to what we imagine the Garden of Eden as being."
- Helmut Lieth (1989). *Tropical Rain Forest Ecosystems: Biogeographical and Ecological Studies*, p.611 "Important food plants, such as sago-producing palms, fruit-producing trees and medicinal plants were purposefully aggregated and tended in convenient places. Eventually, the forest garden, a kind of Garden of Eden, emerged. These jungle gardens on good soils of easy access required little maintenance and hardly any hard work."
- Dave Jacke and Eric Toensmeier (2005). *Edible Forest Gardens - Volume One*, p.1
- Robert Hart (1996). *Forest Gardening: Cultivating and Edible Landscape*, p.80

[5] Patrick Whitefield (2002). *How to Make a Forest Garden*. p. 5.

[6] Robert Hart (1996). *Forest Gardening*. p. 45.

[7] Robert Hart (1996). *Forest Gardening*. pp. 28 and 43.

[8] "Introduction to Permaculture - Compatibility with Veganic Agriculture". Veganic Agriculture Network.

[9] Zalan Glen (2009). "From Permaculture to Vegaculture" (PDF). The Movement for Compassionate Living - New Leaves (issue no.93): 18–20.

[10] Dalziel O'Brien, Kenneth, *Veganic Gardening*, 1986, page 9

[11] *Veganic Gardening*, Kenneth Dalziel O'Brien, page 16

[12] http://www.telegraph.co.uk/news/obituaries/9146710/Maria-Thun.html

[13] Maria Thun: Bäume, Hölzer und Planeten. 2nd edition (2008). page 144 - 146. Note: this chapter was published the first time in the second edition, it cannot be found in the first edition. Unfortunately, there is no english translation, but this book contains a number of useful photographs so it might still be worth it.

[14] http://www.prnewswire.com/news-releases/biodynamics-without-the-cow-horn-171019541.html

[15] http://www.veganitaly.com/

[16] "North Carolina Cooperative Extension, Soil Preparation".

26.5 References and further reading

- *Growing Green: Organic Techniques for a Sustainable Future* by Jenny Hall and Iain Tolhurst. Vegan Organic Network publishing, 2006, ISBN 0-9552225-0-8. Available in the US from Chelsea Green Publishing, ISBN 978-1-933392-49-3.

- *Growing Our Own: A Guide to Vegan Gardening* by Kathleen Jannaway. Movement for Compassionate Living publishing, 1987. ASIN B001OQ7G8S.

- *Plants for a Future: Edible and Useful Plants for a Healthier World* by Ken Fern. Hampshire: Permanent Publications, 1997. ISBN 1-85623-011-2.

- *Veganic Gardening- The Alternative System for Healthier Crops* by Kenneth Dalziel O'Brien. Thorsons Publishing, 1986, ISBN 0-7225-1208-2.

Chapter 27

Village Community Co-operative

The **Village Community Co-operative** was established in the City of Prospect in the late 1970s. The co-operative was established as a multiple occupancy permaculture project at Kuipto Forest, south of Meadows. The community held fairs and meetings in town, field days and workshops at Kuipto and in Prospect.[1]

The Co-operative was originally run from a house on Braund Rd, Prospect. The house had a shop front which was opened as The Village Store. This evolved into Village Natural Technology Systems with a focus on solar, wind, sustainable house design, permaculture, books and workshops.

Village Natural Technology Systems moved to Prospect Road, Prospect in the early 1980s and the Co-operative owned store was opened by the Honorable John Bannon, premier of South Australia, on November 26, 1983 [2]

The Village Community Co-operative was also active in the anti-nuclear movement collaborating with CANE and Friends of the Earth to produce Your carry anywhere anti - uranium songbook for Australia.[3]

27.1 References

[1] Ball, Colin. "Permaculture's First Years in South Australia" (PDF).

[2] Bannon, John (1983), *Natural Technology Systems new premises*, retrieved 3 July 2016

[3] "SASS Record C4594". *www.library.unisa.edu.au*. Retrieved 2016-11-16.

Chapter 28

Village Homes

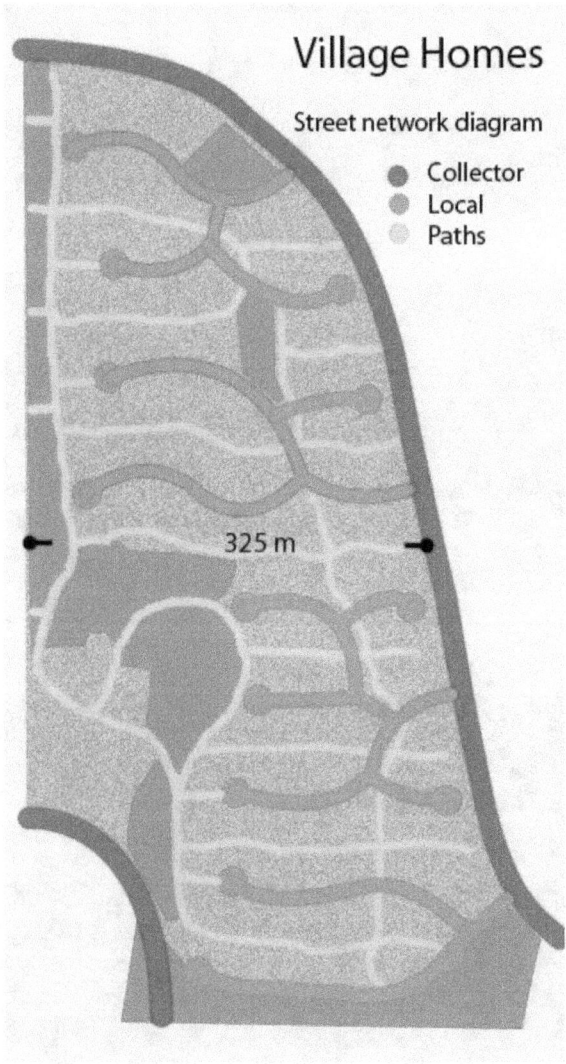

A diagram showing the street and path network of Village Homes.

Village Homes is a planned community in Davis, California, Yolo County, designed to be ecologically sustainable by harnessing the energies and natural resources that exists in the landscape, especially stormwater and solar energy.

28.1   History

The principal designer of Village Homes was architect Mike Corbett who began planning in the 1960s, with construction continuing from south to north from the 1970s through the 1980s. Village Homes was completed in 1982, and has attracted international attention from its inception as an early model of an environmentally friendly housing development, including a visit from then-French President François Mitterrand.[1]

28.2   Contributions to Sustainability

Grass lined swale collects rainwater, which then slowly percolates into the soil where it is protected from runoff and evaporation.

The 225 homes and 20 apartment units that now constitute the Village Homes community utilize solar panels for heating, and are oriented around common areas at the rear of the buildings, rather than around the street at the front. All streets are oriented east-west, with all lots positioned north-south. This feature has become standard practice in Davis and elsewhere, since it enables homes with passive solar designs to make full use of the sun's energy throughout the year. The development also uses natural drainage, called bioswales, to collect water to irrigate the common areas and support the cultivation of edible foods, such as nut and fruit trees and vegetables for consumption by residents, without incurring the cost of using treated municipal water.[2]

28.3   References

[1] Source: About Village Homes; "Archived copy". Archived from the original on June 9, 2007. Retrieved June 13, 2007.; Smart Communities, "Village Homes." Archived August 17, 2007, at the Wayback Machine.

[2] Source: Village Homes, Davis, California; Bill Browning and Kim Hamilton, "Village Homes: A model solar community proves its worth"; Smart Communities, Ibid.

28.4   External links

- http://www.villagehomesdavis.org/

- Village Homes information on the Davis Wiki.

- http://www.context.org/ICLIB/IC35/Browning.htm

- https://web.archive.org/web/20070609124520/http://arch.ced.berkeley.edu/vitalsigns/workup/siegel_house/vh_bkgd.html

- http://web.archive.org/web/20131110035906/http://www.earthfuture.com/community/villagehomes.asp

28.4.1   Documentary videos about Village Homes

- Global Gardener: Urban Permaculture feat. Bill Mollison (1991)

- GeoffLawton.com: Food Forest Suburb (2015)

Coordinates: 38°32′50.92″N   121°46′49.83″W   /
38.5474778°N 121.7805083°W

Chapter 29

Waru Waru

Waru Waru is an agricultural technique developed by pre Hispanic peoples in the Andes region of South America, from Colombia to Bolivia.[1] It is dated around 300 B.C.[2]

The technique has been revived in 1984, in Tiawanaco, Bolivia, and Puno, Peru.

The technique consists in combining raised beds with irrigation channels so as to prevent damage due to soil erosion during floods. The technique ensures both collecting of water (either fluvial water, rainwater or phreatic water) and subsequent drainage. The drainage aspect makes it particularly interesting for many areas subjected to risks of brutal floods, such as tropical parts of Bolivia and Peru where it emerged. Waru waru has been used in many countries, like China.

29.1 See also

- Chinampa

29.2 References

[1] Nivedita Khandekar; Geoffrey Kamadi; Dan Collyns (19 August 2015). "The three wonders of the ancient world solving modern water problems In Peru, Kenya and India, NGOs are helping communities overcome water scarcity using wisdom from the past". The Guardian. Retrieved 19 August 2015.

[2] "Raised beds and waru waru cultivation". Organization of American States, Department of Sustainable Development. Retrieved 2008-03-06.

Coordinates: 15°47′01″S 70°01′41″W / 15.78361°S 70.02806°W

Chapter 30

Paul Yeboah

Paul Yeboah, is an educator, farmer, permaculturist, community developer, and social entrepreneur. He is the founder and coordinator of the Ghana Permaculture Institute and Network in Techiman, Ghana, West Africa. It is located in the Brong-Ahafo Region of Ghana. The purpose of the Institute is to build and maintain a stable food system, to take care of the local ecosystems, and to improve the quality of life in the rural areas.[1][2] The GPN trains students and community in sustainable ecological farming techniques. They support projects through out Ghana; women groups, micro-finance projects; teach growing moringa; mushroom production; alley cropping, food forests development and Agroforestry.[3]

Permaculture is based on natural sustainable design systems. A agricultural system that uses practices to keep soil fertile, crops and livestock healthy. It encourages protection of the environment and an environmental lifestyle; so as to maintain environmental stability and maintain environmental resources for the future. It rehabilitates eroded and deforested land. The Permaculture Network encourages the practice of permaculture at home. The Permaculture Network's mission is to encourage, educate, and promote the use of permaculture by farmers and people in Ghana, which will contribute to the environmental soundness, and stability of the country's future.

They host international volunteers, interns, and students.[4] The Ghana Permaculture Network and Institute is a member of the Ghana Ecovillage Network. Which is an organization of sustainable development leaders and projects. Paul Yeboah is Vice President of the GEN which works towards promoting Indigenous Initiatives and Sustainability in Ghana. Permaculture is transforming communities in Ghana through education, food production, outreach, skills development, self-sufficiency, and creating small business enterprises.[5][6]

Ghana Brong Ahafo Region

30.1 Background

At the age of 22 Paul Yeboah was concerned with rural and urban poverty. He received an Agricultural Certificate from the Farm Institute in Ghana. He initiated a rural and urban fruit forest project by using seed supplies from the Kade Oil Palm Research Institute and Bonsu Cocoa Research Station. The seeds and seedlings were given to the farmers on a credit basis. This project was instrumental in the creation of rural processing businesses and employment for the poor.

In 2003 Paul Yeboah was the farm manager for the Abbott of Kristo Buase Benedictine Monastery in Ghana. Greg Knibbs was invited to come to the Monastery to assist in the redesigning of the farm using Permaculture practices to restore the soil to fertility. The soil was depleted from the use of synthetic chemical pollution. Yeboah met Greg Knibbs and they worked together to form the Ghana Permaculture Network which later became the Ghana Permaculture Institute.

30.2 Career

The Ghana Permaculture Network was coordinated by Paul Yeboah in 2003. The GPN started out as a small farm demonstration training site that later grew into the Ghana Permaculture Institute (GPI). In 2007 the Ghana Permaculture Nwodua Tree Nursery was created. It was created to build community income, and to deal with environmental issues such as desertification and erosion. It also does reforestation. The work is collaborative community effort involving women, youth, and men. All members of the community are a part of the profits and benefits of working towards enhancing their environment. The project builds environmental awareness.[7]

Paul Yeboah is the Vice Chairman of the Ghana Ecovillage Network which was founded in 2012, and incorporated in 2013. It was formed by community leaders and groups with sustainable projects to promote Ecovillage strategies as models for sustainable development in Ghana.[8]

The Ghana Permaculture Network started out supporting local schools, community farmers in establishing tree nurseries, and tree planting projects. The Ghana Permaculture Network has now expanded in various parts of Ghana, into Togo, and Burkina Faso in West Africa.

A photograph of Tree Nursery in Techiman, Ghana.

30.3 Permaculture Institute Projects (Educational Programs)

The goal of the educational initiatives and projects is to promote lifelong learning with emphases on vocational training, and the need of viable skill development using practices of sustainable development.

- Permaculture Design Courses are taught in Ghana, Togo and Burkina Faso, West Africa.

- Ghana Permaculture Nwodua Tree Nursery Project - Ghana is a deforested country. The Tree Nursery Project addresses issues of deforestation in order to reverse erosion. It teaches and advocates the importance of trees and climate change issues.

- Oyster mushroom production - Using sawdust in bags the Permaculture Institute teaches how to grow mushrooms. This project allows participants to learn how to make a waste product sawdust useful. Which promotes healthy diet, nutrition and skill development.

- Demonstration Training sites for how-to build Permaculture home gardens

- Moringa Production - Which helps to facilitate the development of small-businesses to generate income. A project that promotes healthy nutrition and skills development.

- Ecovillage Design Course - Which teaches and acquaints students with tools that can be utilized to redevelop their communities in ecological economically, culturally and socially ways to foster sustainability.

Ghana Permaculture Network - Mushroom Production

Mushroom Vendor, Ghana

30.4   External links

- Paul Yeboah Permaculture Institute

- Permaculture Design Course, Techiman Ghana

- Permaculture Design Course (Part two)

- Permaculture (What is Permaculture?)

- Ghana Permaculture Video Clips

- Permaculture and mushroom cultivation in Ghana

- Adele Women Association 'Upper Volta' Region Ghana

- Gaviotas is a model village of sustainable development in Colombia, South America.

- Urban Sustainable Development in Curitiba, Columbia

- Permaculture Institute - Alley Cropping and Food Forest 0001

- Paul Yeboah on Panel Discussion about Permaculture.

- Permaculture and mushroom cultivation in Ghana

- Future In Our Hands International Network

30.5   Further Reading

- http://www.agriculturesnetwork.org/resources/extra/final-report-de-schutter

- Permaculture in Africa

- UN: Eco-Farming Feeds the World

30.6   References

[1] https://permacultureghana.wordpress.com/what-is-permaculture-2/

[2] http://edge5.com.au/permaculture-courses-fremantle/

[3] http://fiohnetwork.org/wp-content/uploads/2015/09/PAUL-TESTIMONY.pdf

[4] http://www.ghanaweb.com/GhanaHomePage/economy/artikel.php?ID=123567

[5] http://gen.ecovillage.org/en/gen-ghana

[6] https://www.permaculture.co.uk/how-permaculture-is-transforming-ghana

[7] http://permaculturenews.org/2011/11/01/the-ghana-permaculture-nwodua-tree-nursery-project-saving-lives-granting-li

[8] http://gen.ecovillage.org/en/gen-ghana

Chapter 31

List of permaculture projects

Robert Hart's forest garden in Shropshire, England

A **permaculture project** is a deployment of permaculture practices on an ongoing basis.

31.1 Africa

31.1.1 Ethiopia

The Biofarming approach applied in Ethiopia has very similar features and can be considered permaculture. It is mainly promoted by the non-governmental organisation BEA, based in Addis Ababa.

31.1.2 Southern Africa

The UN High Commissioner for Refugees (UNHCR) has produced a report on using permaculture in refugee situations after successful use in camps in Southern Africa and Republic of Macedonia.

31.1.3 Zimbabwe

There are 60 schools designed using permaculture, with a national team working within the schools' curriculum development unit.

31.2 Oceania

31.2.1 Australia

The development of permaculture co-founder David Holmgren's home plot at Melliodora, Central Victoria, has been well documented. Geoff Lawton's Zaytuna Farm next to The Channon in northern NSW, Australia, is a 66-acre medium-farm scale example of permaculture implementation. It is the home base for the Permaculture Research Institute. Begun in 2001, the site is off-grid, and has multiple food forest systems, animal systems, kitchen garden and main crop areas, a large network of water-harvesting earthworks for passive hydration of the site, composting toilets, rocket stove powered showers, straw bale natural buildings, etc.[1]

City Farm Perth is an example of community permaculture in an inner suburb of Perth, Western Australia. The farm was constructed on a brownfield site in 1994, and is a focal point for permaculture education, as well as community music and art.[2]

31.3   Asia and the Middle East

Permaculture Institute Asia [3] (PIA) lists most all major permaculture projects and sites in the Asia region. Including Cambodia, China, India, Indonesia, Israel, Palestine, Japan, Korea, Kyrgyzstan, Lao, Malaysia, Micronesia, Nepal, Mongolia, Pakistan, Philippines, Sri Lanka, Tiawan, Vietnam, Brunei, Indonesia, Myanmar, Bhutan

31.3.1   India

Living Ecology and Permaculture Patashala provide higher education to permaculture professionals through practical permaculture projects in small villages. The drylands projects help poor rural farmers with sustainable agriculture and permaculture design implementation.[4][5]

31.3.2   Thailand

Permaculture Institute Thailand (PIT) is an institute representing Permaculture projects in Thailand. International courses and residencies as well as Natural building, aquaculture, Food forestry, vermiculture and many more courses and skills in self-sufficiency.[6]

The Phayao Permaculture Center, Phayao, Northern Thailand, is a permaculture project demonstrating permaculture design including education open to all interested in permaculture.

The Permaculture Research Institute Asia (PIA), located in Buri Ram Province, Thailand, is pined on Google maps, it is a permaculture demonstration and education site. Courses and demonstration of permaculture design techniques, bio intensive gardening and farming methods and regenerative ecosystem designing. Integrated aquaculture and ducks with rice paddy, they show how higher diversity leads to higher yields and food security which can be obtained from rice fields/paddy in the Mekong Bio region.

31.3.3   Indonesia

Bumi Langit Institute is an initiative representing Permaculture projects in Indonesia, resided in Yogyakarta. The Institute founded by Iskandar Waworuntu Alhajj an environmentalist veteran in 2006. International and national courses and residencies as well as natural building, aquaculture, food forestry, vermiculture, bamboo culture, coconut culture and many more courses and skills in self-sufficiency, regenerative design and leadership.

31.3.4   Cambodia

A consortium of NGOs including Lom Orng[7] and Ockenden[8] is doing a post-flood livelihood and infrastructure regeneration project, in the country's northwest, which includes permaculture principles, and the establishment of a permaculture demonstration farm in Battambang Province which serves as a community farm and education site and includes a native tree nursery and biogas system providing clean cooking fuel and lighting.[9]

A Facebook Premaculture Cambodia page list various Permaculutre projects in the Cambodia region and the proprieters.

31.3.5   Nepal

The Himalayan Permaculture Centre (HPC) is a grass roots non-government organisation (NGO) set up by trained and motivated farmers from Surkhet district (Mid-Western Nepal) in 2010 to implement sustainable rural development programs in Nepal.[10]

31.3.6   Saudi Arabia

The Al Baydha Project, a land restoration and rural community development program in Western Saudi Arabia.

31.4   Europe

31.4.1   Cyprus

GROL Garden — Girne: An Urban Permaculture Project that accepts Volunteers who come under a work-exchange agreement. GROL Garden provides food, accommodation, yoga classes, and hands-on learning in exchange for assistance on their Projects.[11][12][13]

31.4.2   Spain

Red de Permacultura Ibérica (Iberic Permaculture Network)[14]

31.4.3   Portugal

In Portugal there are several Permaculture projects. There's a bottom-up initiative that is mapping most of them called Rede CONVERGIR.[15]

Tribodar Learning Center, located in the center of Portugal, provides an holistic learning experience that allows those involved in the learning process to gain growth on different levels. Having the freedom to learn at our own pace is another element of this learning approach.[16]

31.4.4 Romania

- Gradina din Gura Siriului is an urban Permaculture initiative located in the periphery of Bucharest. The project combines permaculture with arts and community events also provides free camping space for travelers all over the world.

31.4.5 United Kingdom

There are a number of example permaculture projects in the UK, including:

- Agroforestry Research Trust, managed by Martin Crawford, is a not-for-profit organisation based in Dartington, Devon that runs a 2-acre (8,100 m^2) forest garden and publishes the journal Agroforestry News [17]

- Chickenshack Housing co-op,[18] a fully mutual housing co-op established in 1995 using permaculture design principles. Based in rural North Wales, the community has 4 dwellings and 6 residents on a 5-acre (20,000 m^2) site. Features include a biomass and solar district heating scheme, a half-acre forest garden and various wildlife conservation and habitat creation strategies. The community is very active in regional sustainability projects such as the Machynlleth Transition Towns initiative. It runs occasional courses in permaculture design and regularly receives visits from interested parties.

- Middlewood Trust, a permaculture-based farm in North Lancashire running courses in permaculture, crafts, forestry and sustainability[19]

- Plan-It Earth Eco Project offers a variety of environmental education programs, courses and events, and eco-holidays at a traditional smallholding near Penzance, Cornwall[20]

- Plants for a Future, a vegan-organic project based at Lostwithiel in Cornwall that is researching and trialing edible and otherwise useful plant crops for sustainable cultivation. Their online database features over 7,000 such species that can be grown within the UK.[21] A collaborative version of the database is in development by the permaculture.info project.

- *Prickly Nut Woods*, a 10-acre (40,000 m^2) woodland near Haslemere, Surrey that is managed by Ben Law. He uses a 'whole system' permacultural approach, using a wide variety of woodland products and documenting a complex web of relationships. He built a house almost entirely using products from the woodland, which was featured in Channel 4's Grand Designs TV series.[22]

- Ragmans Lane, a 60-acre (24 ha) farm in the Forest of Dean in Gloucestershire.[23]

- The RISC Roof Garden, on top of a development education centre in Reading city centre and inspired by Robert Hart's permaculture forest garden in Shropshire, is an excellent example of urban permaculture design.[24] It is used by schools, educators and designers as an educational resource for sustainable development and is a member of the National Gardens Scheme. The garden is composed of dense plantings of over 180 species of edible and medicinal plants and is fed by rainwater and composted waste from the centre.

- Tir Penrhos Isaf, near Dolgellau, developed by Chris and Lyn Dixon since 1986.[25]

- The Sussex Roots Society, a community permaculture garden initiated by Mischa Nowicki and Integralpermanence at Sussex University, collaborating with the student union as part of the countries first "Eco- University" initiative.

- Eastside Roots- is a (not for profit, worker co-op) community garden centre and hub which evolved from Bristol Permaculture Group. 'Waste' ground owned by First Great Western has been leased and is a garden centre with ongoing education, social projects and outreach into the wider community. The project promotes sustainability and permaculture.

Other projects tend to be more community oriented, particularly in urban areas. These include Naturewise, a north London based group that tends a number of forest gardens and allotments as well as running regular permaculture introductory and design courses;[26] and Organiclea, a workers cooperative that is involved in developing local food-growing and distribution initiatives around the Walthamstow area of east London.[27] The Transition Towns movement initiated in Totnes and Kinsale by Rob Hopkins is underpinned by permaculture design principles in its attempts to visualise sustainable communities beyond peak oil.[28]

The UK Permaculture Association publishes an extensive directory of other projects and example sites throughout the country.

31.5   North America

The Permaculture Association of Teachers and Organizers on WiserEarth maintains a US listing.[29]

31.5.1   Northeast US

- Prospect Rock Permaculture is one of the region's longest established permaculture education and demonstration sites, and Vermont's oldest permaculture design firm. They grow a diversity of nuts, fruit trees, vines, berries, vegetables, medicinal and culinary herbs and teas, produce eggs and honey, and host a variety of workshops, permaculture courses, advanced permaculture design internships, and other educational programs.[30]

- The Northeastern Permaculture Network brings together enthusiasts in the northeastern United States and eastern Canada.[31] The Northeastern Permaculture Convergence has been held every summer in a different location in the northeast. There is also the 501(c)3 P.I.N.E (Permaculture Institute of the Northeast).[32]

- People's Garden Pittsburgh is building a one-acre food forest on an abandoned landslide in the Hill District of Pittsburgh, which is considered a food desert.[33][34]

- Sowing Solutions Permaculture Design & Education offers a Permaculture Design Certification Course twice a year in Western MA, either as a twelve-day summer intensive course, or as a weekend series in the spring/ fall. The course includes visits to numerous permaculture demonstration sites in the region, and guides students through the ecological design process.

- UMass Permaculture <ref="UMass">"UMass Permaculture | UMass Dining". *umassdining.com*. 2013. Retrieved June 17, 2013.</ref> is a university permaculture initiative that transforms marginalized landscapes on the campus into diverse, educational, low-maintenance and edible gardens, according to UMass officials.[35] During November, 2010, "about a quarter of a million pounds of organic matter was moved by hand",[36] using all student and community volunteer labor and no fossil fuels on-site. The process took about two weeks to complete. Now, the Franklin Permaculture Garden includes a diverse mixture of "vegetables, fruit trees, berry bushes, culinary herbs and... flowers that attract beneficial insects."[36]

- Burlington Permaculture[37] is an ad-hoc community group based out of Burlington Vermont. BTP unites neighbors to promote urban agriculture and reforestation, enhance neighborhoods, and strengthen the web of community resources as we look beyond sustainability towards a vibrant healthy relationship with our landscape. The group was founded by Michael Blazewicz of Round River Design, Mark Krawczyk of Rivenwood Crafts, and Keith Morris of Prospect Rock Permaculture.

- Green Phoenix Permaculture,[38] offers design courses and workshops, demonstrating forest gardens, natural building, permaculture designs, and a community of many permaculture designers & teachers including AppleSeed Permaculture[39] and Sowing Solutions Permaculture Design & Education.[40]

- The websites FLXpermaculture.Net[41] and the Upstate NY Permaculture Network[42] are email lists and web sites for permaculture practitioners in New York state.

- The Finger Lakes Permaculture Institute[43] offers an annual Permaculture Design Certification course, apprenticeships, and workshops that are often held at area sites designed using permaculture principles. The Finger Lakes Permaculture Institute was founded by Michael Burns, Steve Gabriel, and Karryn Olson Ramanujan in 2005.

31.5.2   Southeast US

- Shades of Green Permaculture Design, Inc., Atlanta, GA designs and builds ecological landscapes that provide organic food, medicine, beauty, fertile soils, and clean water, while building resilient communities and local economy. In addition to leading a 6-month Permaculture Design Certification course annually, this design firm serves clients ranging in scale from project management of new development, to institution-level research farms, from broad-acre restoration and conservation, to small-scale residential homesteads. In addition to their extensive client sites, they have demonstration gardens based in Pine Lake, GA.

- The Farm Ecovillage Training Center, Summertown, Tn[44] provides examples drawn from 4 decades of ecovillage living, including cob, earthbag, round-pole and strawbale buildings, constructed wastewater wetlands, fish, ducks and chickens, organic and biodynamic gardens, orchards, shiitake production, biofuel production, solar photovoltaics, herbal tinctures and plant medicines, humanure and other composting practices, and community conviviality processes.

- Spiral Ridge Permaculture Gardens, Summertown, Tn[45]

- Cedar House Inn Bed and Breakfast/Permaculture Lifestyles Permaculture Demonstration Project, Dahlonega, Georgia[46]{http://www.permaculturelifestyles.com Permaculture Lifestyles}, Tours given to guests of inn and clients.

- The Edible Plant Project implements and promotes elements of permaculture through a nonprofit nursery and workshops in Gainesville, FL (home of University of Florida).[47]

- Georgia Permaculture Institute, Central Georgia

- Knoxville Permaculture Guild, Knoxville, TN.[48]

- Northern Virginia Permaculture Guild[49]

- Sweet Aspect Permaculture, the steward of one acre of land outside of Asheville, NC, is experimenting with permaculture principles, design, and implementation. Sweet Aspect focuses on small animal management, including lambs and chickens, gardens, orchards, edible forest management, shitake production, rainwater catchment, greywater systems, natural and green building, fermentation, herbal tinctures and plant medicines, humanure and other composting practices, and alternative fuel and energy harvesting.

- The Urban Permaculture Institute of the Southeast in Walterboro, South Carolina practices permaculture on a 1/2 acre lot in the middle of the city with chickens, fish, bees, food forestry, vermiculture, insect production, greywater systems, passive solar, mushroom production, rainwater catchment, floating hydroponics and partial hydroponic systems.

- Spring Village / New Community Project in Harrisonburg, VA[50] practices and teaches permacture at a small urban site.

31.5.3 Midwestern and Rocky Mountain region

- Permaculture Research Institute for Cold Climate.[51]

- The Round Mountain Institute in the Gunnison valley of Colorado is a nonprofit that is dedicated to sustainable agriculture high in the rocky mountains near the continental divide.

- Central Rocky Mountain Permaculture Institute.[52]

- The Urban-Suburban Sustainability Initiative based in Belleville, Illinois is a local grassroots organization in the process of starting up. Its focus will be on permaculture, bioremedification, environmental education and Local Exchange Trading Systems.

- Bloomington Permaculture Guild, Bloomington, Indiana

- High Altitude Permaculture Institute is based in Ward, Colorado, which is west of Boulder.[53]

- Midwest Permaculture[54] delivers up to a dozen PDC Courses per year (in the Midwest and elsewhere) along with many other kinds of permaculture trainings.

- The Center for Sustainable Community[55] asked Midwest Permaculture to do a permaculture design for the 8.7 acres of land that immediately adjoins the local community in Stelle, Illinois.[56]

- Bur Oak Farm in SE Wisconsin, approximately 40 miles north of Milwaukee

- Urban Permaculture Design a city lot with over a hundred perennial edible varieties. Permaculture land acquisition discussion. Urban Classes.

- Sunrise Ranch Intentional Community near Loveland, Colorado. Farm to table internship programs, permaculture design course, farm/garden internships.

31.6 Latin America

31.6.1 Belize

Maya Mountain Research Farm in San Pedro Columbia, Toledo, Belize, is a permaculture project founded in 1988 by Christopher Nesbitt on a 70-acre abandoned citrus and cattle land. The land is up the Columbia Branch of Rio Grande from the Kekchi maya community. Presently the farm manages over 500 species of plants. The farm works on agroecology, permaculture, stacked polycultures and installs photovoltaic systems in schools in indigenous communities as well as protected areas.

The farm works closely with farmers organizations and holds one Permaculture Design Course per year.

31.6.2 Costa Rica

Saint Michaels Sustainable Community started building their Permaculture Ranch in 2001 in the mountains of Esterillos, Costa Rica 3.5 Kilometers from the beach. By 2015 they have become internationally recognized model of regenerative agriculture. The building of the community was accomplished by interns, volunteers, woofers, ecological experts, a NASA scientist,agriculture students, permaculturists, local farmers, and environmentalists (some that formerly worked for oil companies)and a regenerative rancher, Justin Dolan. Under Dolans' leadership five living roof homes have been designed and built using permaculture principals and two other homes have been built with rainwater harvesting roofs. A natural salt water swimming pools was installed in a food forest containing over hundreds of types of fruits, medicinal herbs, a free energy aquaponics system, 12 rotational pastures, chickens, lambs, a herd of cattle, reforestation, an equestrian center, organic orchards, and a disc golf course. The Community is home to the Permaculture Country Club of Costa Rica (PCC). The club teaches a Journeyman Apprentice Program. The Candidates need to demonstrate expert knowledge and experience of satisfactory performing the duties involved from start to finish of each area of Permaculute Namely:

Aquaponics design, feeding, harvesting of the fish, inputs and outputs Hugleculture BioChar Swales: building design, planting Edible Landscaping Terrace gardening Rotation of crops and gardens Cover crops Keyline design Basic Organic animal care Rotational Grazing Organic orchard management Introduction to medicinal plants and holistic animal care Driving cattle Agroforestery Food forests Energy efficiency Check dams Plant propagation Planting seedlings, transplanting seedlings, planting trees in square holes, Pruning closer together and pruned in for less water and higher yield Harvesting Building garden beds Gilds Zones Making compost tea; Higher recipe for fruit trees Weed identification, weeding Rain water harvesting Composting methods Natural Building Soil regeneration and ecology Patterns and edge planting Holistic management Bio Dynamics Sustainably intrigued with living Living roof design and construction Twelve design principles of Permaculture Zero Waste / Upcycling/ recycling - Producing no waste And Disc golf [57]

31.6.3 Guatemala

The Instituto Mesoamericano de Permacultura (IMAP)[58] was founded in 2000 by a group of Mayan farmers and professionals in San Lucas Toliman, Guatemala. It is a not-for-profit community organization focused on the development of self-sufficient communities through the responsible management of natural resources, using permaculture techniques and ancestral and traditional knowledge. Areas of focus include: the cultural and biological diversity of Mesoamerica; food security and food sovereignty; organic production systems at the community and familial level; and permaculture education and environmental stewardship.

31.6.4 Cuba

Main article: Organoponicos

Since 1993, Cuba has transformed its food production using low-input, or organic agriculture and, to some degree, permaculture. Havana produces up to 50% of its food requirements within the city limits, much of it organic and produced by people in their homes, gardens and in municipal spaces[59] The transformation in agriculture originated as a response to a crisis which is known as the Special Period. Starting in 1993 Australian permaculturists traveled to Cuba to educate Cuban gardeners in permaculture practices.[60][61]

31.6.5 Ecuador

- Guaycuyacu http://www.guaycuyacu.net/ - in the foothills of northwestern Ecuador

- Black Sheep Inn http://blacksheepinn.com/ - Chugchilán, Cotopaxi, Ecuador

- Bosque de pas / Bosque de bambu http://www.bospas.org/ - Ibarra, Ecuador

- Finca mono verde http://www.fincamonoverde.com/ - Tabuga, Ecuador

- Finca la amistad http://www.fincalaamistad.net/ - Imbaruba, Ecuador

- Finca Sagrada http://www.fincasagrada.net/ - Vilcabamba, Ecuador

- Rio Muchacho http://www.riomuchacho.com/ - Canoa, Ecuador

- Neverland Farm http://www.neverlandfarm.org/ - Vilcabamba, Ecuador

- Sacred Suenos https://sacredsuenos.wordpress.com/ - Vilcabamba, Ecuador

- Eagle condor Farm http://www.eaglecondorfarm.com/ - Alausi, Ecuador

- Finca verde https://fincaverde.wordpress.com/ - Valladolid and Palanda, in Zamora-Chinchipe, Ecuador

- Terra Frutis http://www.terrafrutis.com/ Gualaquiza, Ecuador

- Finca Ave Terra - mindoregenerativefarm - Mindo, Ecuador

31.7 References

[1] "Geoff Lawton's Zaytuna Farm Video Tour (Apr/May 2012) - Ten Years of (R)Evolutionary Design Permaculture Research Institute — Permaculture Forums, Courses, Information & News". *permaculturenews.org*. 2013. Retrieved June 17, 2013.

[2] "Perth City Farm". *perthcityfarm.org.au*. 2013. Retrieved June 17, 2013.

[3] http://www.PermacultureInstituteAsia.com

[4] http://www.livingecology.org

[5] permaculturepatashala.com

[6] http://www.PermacultureInstituteThailand.org

[7] "LOM ORNG ORGANISATION: Cambodian War Amputees and Disaster Warning and Prevention". *lomorng.org*. 2013. Retrieved June 17, 2013.

[8] "Welcome to Ockenden Cambodia". *ockendencambodia.org*. 2009. Retrieved 17 June 2013.

[9] "Permaculture Demonstration Farm with Biogas | Evolution in Motion". *evolutioninmotion.org*. 2013. Retrieved June 17, 2013.

[10] "Welcome — Himalayan Permaculture". *himalayanpermaculture.com*. 2013. Retrieved June 17, 2013.

[11] http://www.GROLgarden.info

[12] "Help at an ORGANIC Garden in North Cyprus". *workaway.info*.

[13] "The Permaculture Research Institute of Australia". *permacultureglobal.com*.

[14] "Red de Permacultura Ibérica / FrontPage". *reddepermaculturaiberica.pbworks.com*. 2013. Retrieved June 17, 2013.

[15] http://www.redeconvergir.net

[16] Ecovillage, TLC (2015). "Ecovillage - Tribodar Learning Center". *Ecovillage*. Retrieved 30 May 2015.

[17] "Agroforestry research trust fruits nuts seeds plants publications". *agroforestry.co.uk*. 2013. Retrieved June 17, 2013.

[18] "Chickenshack Housing Co-operative Limited". *chickenshack.co.uk*. 2010. Retrieved June 17, 2013.

[19] "Middlewood Trust". *middlewood.org.uk*. 2013. Retrieved June 17, 2013.

[20] "Plan-it Earth, Sancreed, Cornwall — Home". *planitearth.org.uk*. 2013. Retrieved June 17, 2013.

[21] "Plants For A Future : 7000 Edible, Medicinal & Useful Plants". *pfaf.org*. 2013. Retrieved June 17, 2013.

[22] "A Visit to Ben Law's Woodland House | Permaculture Magazine". *permaculture.co.uk*. 2013. Retrieved June 17, 2013.

[23] "Home". *ragmans.co.uk*. 2013. Retrieved June 17, 2013.

[24] "Gardens | risc.org.uk". *risc.org.uk*. 2013. Retrieved June 17, 2013.

[25] "penrhos home page permaculture design and sustainable development". *konsk.co.uk*.

[26] The Naturewise Forest Garden

[27] "- OrganicLea – A workers' cooperative growing food on London's edge in the Lea Valley". *OrganicLea — A workers' cooperative growing food on London's edge in the Lea Valley*.

[28] Rob Hopkins. "Why 'Transition Culture'? » Transition Culture". *transitionculture.org*.

[29] http://www.wiserearth.org/group/PATO

[30] "Prospect Rock Permaculture | Design and Education for Ecological Culture". *prospectrockpermaculture.wordpress.com*. 2013. Retrieved June 27, 2013.

[31] "Northeastern Permaculture Networ — home". *northeasternpermaculture.wikispaces.com*. 2013. Retrieved June 17, 2013.

[32] "P.I.N.E.". *thepine.org*. 2012. Retrieved June 27, 2013.

[33] "The People's Garden, Pittsburgh — Home". *peoplesgardenpittsburgh.com*. 2013. Retrieved June 27, 2013.

[34] "'Permaculture Center for BioRegional Living | Andrew Faust Design Consultation.'". *homebiome.com*. 2013. Retrieved June 17, 2013.

[35] Permaculture garden at UMass gives new meaning to the phrase fresh vegetables *The Daily Hampshire Gazzette*

[36] UMass Embraces Permaculture *Food Service Director*

[37] "burlingtonpermaculture2". *sites.google.com*. 2013. Retrieved June 17, 2013.

[38] "Green Phoenix — Home". *green-phoenix.org*. 2013. Retrieved June 17, 2013.

[39] "AppleSeed Permaculture: Edible Landscaping & Regenerative Design". *appleseedpermaculture.com*. 2013. Retrieved June 17, 2013.

[40] "Sowing Solutions — Sowing Seeds for Ecological Living". *sowingsolutions.net*. 2013. Retrieved June 17, 2013.

[41] "Finger Lakes Permaculture Network | FLXpermaculture.Net". *flxpermaculture.net*. 2013. Retrieved June 17, 2013.

[42] "Upstate NY Permaculture Network". *upstatenypermaculture.net*. 2013. Retrieved June 17, 2013.

[43] "Finger Lakes Permaculture Institute". *fingerlakesperma-culture.org*. 2013. Retrieved June 17, 2013.

[44] "The Farm Ecovillage Training Center ← Providing train-ings and apprenticeships in Ecovillage Design, Research, Implementation & Education". *thefarm.org*. 2013. Re-trieved November 20, 2013.

[45] "Spiral Ridge Permaculture ← Specializing in Agro-Ecosystems Design, Research, Implementation & Educa-tion". *spiralridgepermaculture.com*. 2013. Retrieved June 17, 2013.

[46] "Demonstration Site". *permaculturelifestyles.com*. 2013. Retrieved June 17, 2013.

[47] "Edible Plant Project". *edibleplantproject.org*. 2013. Re-trieved June 17, 2013.

[48] "Knoxville Permaculture Guild — Building Permanent Cul-ture in Knoxville". *knoxvillepermacultureguild.ning.com*. 2013. Retrieved June 17, 2013.

[49] "Northern Virginia Permaculture Guild (Arlington, VA) - Meetup". *meetup.com*. 2013. Retrieved June 17, 2013.

[50] "New Community Project Blog". Retrieved June 4, 2014.

[51] "PRI Cold Climate — PRI Cold Climate Research Urban Planning Ecology Classes". *pricoldclimate.org*.

[52] "Home". *crmpi.org*. 2013. Retrieved June 17, 2013.

[53] "High Altitude Permaculture". *hialtpc.org*.

[54] "Midwest Permaculture". *midwestpermaculture.com*. 2013. Retrieved June 17, 2013.

[55] "CSC Home |". *cscstelle.org*. 2013. Retrieved June 17, 2013.

[56] "The Permaculture Design for CSC in Stelle, IL Midwest Permaculture". *midwestpermaculture.com*. 2013. Retrieved June 17, 2013.

[57] http://www.saintmichaelscostarica.com

[58] "About IMAP | Instituto Mesoamericano de Permacultura [IMAP]". *imapermacultura.wordpress.com*. 2013. Re-trieved June 17, 2013.

[59] "The Origin of Permaculture in Cuba | Sustainable Cities Collective". *sustainablecitiescollective.com*. 2013. Re-trieved June 17, 2013.

[60] Tiller, Adam (1995). "AIDAB/NGO Cooperation Program, Food Gardener Education in Urban Havana, 1995, Final Re-port from the Field". *Australian Conservation Foundation, Cuba*. Retrieved 21 July 2014.

[61] Tiller, Adam (1996). "Información sobre la dirección del proyecto en Cuba: el Green Team.". *Foundation for Na-ture and Humanity's Permaculture Project*. Retrieved 30 June 2014.

31.8   External links

- The Worldwide Permaculture Network
- The Permaculture Research Institute
- The Permaculture Association

31.9 Text and image sources, contributors, and licenses

31.9.1 Text

- **Permaculture** *Source:* https://en.wikipedia.org/wiki/Permaculture?oldid=752138337 *Contributors:* Marj Tiefert, Tarquin, Rmhermen, Pierre-Abbat, Anthere, Tzartzam, BryceHarrington, Quercusrobur, Jose Icaza, DennisDaniels, Infrogmation, Michael Hardy, Dmd3e, Matthewmayer, Cyde, Skysmith, Mac, BigFatBuddha, Scott, Ghewgill, Jengod, Ww, Populus, Jose Ramos, Mignon~enwiki, Vespristiano, Altenmann, Chop-chopwhitey, Flauto Dolce, Gidonb, Sunray, Sheridan, Alan Liefting, DocWatson42, Timpo, Mboverload, JRR Trollkien, Golbez, Geoffspear, Gadfium, Pgan002, Onco p53, Phil Sandifer, Neffk, Zfr, Nickptar, Burschik, Jayjg, Kathar, NathanHurst, Chris j wood, Rich Farmbrough, Guanabot, Vsmith, Dyl, Bender235, Eadmund~enwiki, Erauch, Nigelj, Smalljim, Cmdrjameson, Vortexrealm, Oop, Ziggurat, Timl, Giraffedata, Jkh.gr, Pearle, Mdd, Shafaki, Bmeacham, Paleorthid, Davenbelle, Linmhall, Stillnotelf, Velella, Tony Sidaway, Talkie tim, Blaxthos, Dan100, Rzelnik, RyanGerbil10, Kevin Hayes, FrancisTyers, Cyclotronwiki, Poppafuze, Mindmatrix, RHaworth, Polyparadigm, SP-KP, Jeff3000, Jwanders, Bluemoose, Raines, Palica, Behun, Mandarax, Elvey, Rjwilmsi, Salix alba, Schlüggell, Smithfarm, DoubleBlue, MarnetteD, Jeffmcneill, MikeJ9919, FlaBot, SchuminWeb, Freddydesouza, Jrtayloriv, Monkofthetrueschool, Vmenkov, Roboto de Ajvol, YurikBot, Wavelength, NT-Bot~enwiki, Waitak, JarrahTree, RussBot, TheMoot, Diliff, Pigman, David Woodward, Shell Kinney, Pseudomonas, Dialectric, Mkbnett, Nirvana2013, Kiaparowits, Thesloth, PeterBirkett, Irishguy, RL0919, Epipelagic, TastyCakes, Morgan Leigh, CQ, Meika, Arthur Rubin, Tevildo, Chriswaterguy, Naught101, Mdwyer, Meegs, That Guy, From That Show!, SmackBot, Eclipsenow.org, Sanman nor, Lord Matt, Jtneill, KVDP, Scottlondon, Cacuija, JFHJr, Gilliam, OrionK, Afa86, Schmiteye, Chris the speller, Te24409nsp, Thumperward, Jon513, Salvor, Uthbrian, Colonies Chris, Chendy, Peter Campbell, Sholto Maud, Willow4, Brimba, Neo139, Josh64, JonasRH, Nihilo 01, Djcmackay, Ggpauly, Gurdjieff, Hank chapot, Joli Rouge, Byelf2007, Archimerged, Valfontis, Khazar, SilkTork, Sociotard, Danny Beaudoin, Ckatz, Rkmlai, Beetstra, LuYiSi, WaynaQhapaq, Johnmc, RichardF, Libertyblues, Christian Roess, Nehrams2020, HisSpaceResearch, Iridescent, Ted11, CoulterTM, Mulder416sBot, RookZERO, Ayanoa, IronChris, Grayson wyatt, RiotGearEpsilon, CmdrObot, Tanthalas39, Drinibot, Tahirs, Montanabw, Cydebot, Unclejedd, Paul Millsom, Macropneuma, Daniel J. Leivick, Teratornis, Kozuch, Richhoncho, Trueblood, Thijs!bot, Epbr123, Homohabilis, Daniel, Trevyn, Itsmejudith, Angusscown, Amberckerr, Blathnaid, Kanejamison, Nom DeGuerre, Brian Boyd, Gioto, Luna Santin, Wengero, Tenzicut, Julia Rossi, Adam Chlipala, Papipaul, Lfstevens, Ingolfson, Aquaponics, JAnDbot, Krishvanth, Tomintaz, Barek, Freddy011, Struthious Bandersnatch, Rjholmer, Bdpermie, Roidroid, VoABot II, Appraiser, Steve@sector39.co.uk, APB-CMX, Sustainableyes, Der-Hexer, Edward321, TimidGuy, MartinBot, 4492tues12, Cbuddenhagen, Andre.holzner, Dan arndt, VirtualDelight, UrthBound, GomerMcFlarp, Tgeairn, J.delanoy, Keithkml, MatheoDJ, TaylorAshton, Charlesjustice, Skier Dude, Belovedfreak, Cjstanonis, Madbishop, Jorfer, AprilSKelly, Woodsguy, Scott Roy Atwood, Agerry, Jamesofur, Gracoo2, Inwind, DASonnenfeld, Chrlaney, Dominoconsultant, VolkovBot, Dlesjack, Jwitch, Aesopos, Philip Trueman, TXiKiBoT, Jackovacs, Charlesdrakew, Noformation, Ilyushka88, Wingedsubmariner, Mooreds, Woodlandcreek, Cymon Fjell, Mexeno1, Red58bill, Logan, Richardtelford, Terriemiller, SieBot, Zelchenko, Flyer22 Reborn, Permacultura, Skipsievert, Yone Fernandes, Zentomologist, Lightmouse, Seedbot, Chrisrus, Bodhi Peace, Wetwarexpert, SlackerMom, Sfan00 IMG, ClueBot, Allthingsgreen, Fyyer, The Thing That Should Not Be, Der Golem, LMFernandes, Xavexgoem, Isaebellaspuppetshow, PMDrive1061, 718 Bot, Mynameisnotpj, Ice Cold Beer, Ceilican, Bridgetsgirl, Ecureuil espagnol, Fishnut, PermaculturePlanet, Dana boomer, DumZiBoT, XLinkBot, WikHead, SilvonenBot, Skyeriquelme, Guydavies, Zodon, Luminaia, Ghost accounty, Hunchenfest, Addbot, Wikepermie, TomorrowsDream, ClaireofKLARITY, Some jerk on the Internet, MrOllie, Download, Favonian, SpBot, Granitethighs, 5 albert square, Bluenijin, Ajkoen, Tide rolls, Jarble, Bermicourt, Legobot, Luckas-bot, Yobot, WikiDan61, Themfromspace, Fraggle81, Santryl, Isotelesis, Suvicaya, AnomieBOT, Archon 2488, BrettScott, Materialscientist, Elmmapleoakpine, Citation bot, NinetyNineFennelSeeds, ArthurBot, Xqbot, Miltonics, Apothecia, Kng442, Anna Frodesiak, NathanielGallion, J04n, Katsteele, Kyng, Peaksurfer, Sceloporus, Brambleshire, Rickproser, Ciclotan, FrescoBot, Legion23, Element deck, Augustart, Pentref, Questionthedominantparadigm, Dogposter, Corpuscollosium, Luke831, Lothar von Richthofen, Johnwhol, Zaricki811, Koleszar, DanTheSeeker, Glacier2009, Tutor65, Enloop, BoundaryRider, Quesauth, LAlexanderson, Jan Permaculture, Gelatinouscube42, Mikal42, Gardenlily, Nattydreader, Pat604, Viellashipley, Pinethicket, LittleWink, Smuckola, Yogi tom, H4stings, Kirstendirksen, Darkohead, Steinpal, Lvec-jayson, Permapower, Teatimetrvaelller, Ecoescuela, SweetAspect, Mcalison, Jonkerz, Hauntu2, Permaculture institute, AbeColey, PleaseStand, Tbhotch, TheMesquito, DARTH SIDIOUS 2, Obsidian Soul, RjwilmsiBot, Mukogodo, Stephen.J.Arnold, Grifen2, User2112, Pinqpanther, T3dkjn89q00vl02Cxp1kqs3x7, Look2See1, AbbaIkea2010, Dewritech, Steko, Gagarine, ZéroBot, Elefectoborde, DerekG99, Jeanpetr, Joshfinnie, Southofsouth, Wiooiw, Wayne Slam, Tiago Penedo, Sediks, Yogazeal, Popok75, Nld.rnsm, Rr.nz, Sequoia D, Minuoh, ClueBot NG, Wisdawn, River road permasite, TadewiGomda, Omair00, Eniodros, Krshwunk, Greenman2011, Mesoderm, Widr, Helpful Pixie Bot, ?oygul, Sax66, Charles Gran, Gob Lofa, BG19bot, MKar, Iamharb, Ostrichfern, Northamerica1000, Marcocapelle, Panchito62, Rowan Adams, Willszal, Thepidding, Madboy23, BattyBot, Greyphox, Fairtheewell, Sinique, ChrisGualtieri, Jray310, TheJJJunk, Vckidd, Soransoran, Sminthopsis84, Ghsqueiroz, Permbuddy, Aymankamelwiki, Ngulevski, KcamKcim, Jimkio12, Oioidoug, Jeffersonfranklin, BreakfastJr, HaroldTheHat, BrooklynAve, Amanda Sturgill, Perma2, Mueller felix, Redddbaron, Geoffmen, Will-o-the-west, Presi1980, Noyster, Suavicm, EricEnfermeroMobile, Permaculture design, Mpathfinder, Monkbot, Seandixonsul, James Kern, Ephemeralcas, Trackteur, PermacultureOne, Enzo at Permaculture Education, Jbanegas, Motoindustries, Steveburns888, CV9933, KasparBot, Qf2345, Mcgrubso, Bobsya, Paugoo, Rpnbrn, Aranyagardens, Dutral, Elekes Andor, Adeptmouse, Chickpecking, Bender the Bot, KPPermaculture, Zcarstvnz, AbdulA and Anonymous: 510

- **Agroecological restoration** *Source:* https://en.wikipedia.org/wiki/Agroecological_restoration?oldid=749260423 *Contributors:* Michael Hardy, Cydebot, JaGa, UnCatBot, Vejvančický, Yobot, Johnberrout, Glacier2009, Orenburg1, Jonkerz, Look2See1 and Anonymous: 4

- **Arborloo** *Source:* https://en.wikipedia.org/wiki/Arborloo?oldid=751551565 *Contributors:* Anthony Appleyard, Carbon Caryatid, Bgwhite, SmackBot, George100, SummerPhD, Magioladitis, Fabrictramp, Mufka, Phil Bridger, Tatterfly, Addbot, LinkFA-Bot, Emmanuel.boutet, Xqbot, PatternSpider, Lotje, ArwinJ, ClueBot NG, Helpful Pixie Bot, BG19bot, Daddypawid, Dexbot, EvMsmile, EChastain and Anonymous: 4

- **Companion planting** *Source:* https://en.wikipedia.org/wiki/Companion_planting?oldid=752048083 *Contributors:* The Anome, Tarquin, Heron, Quercusrobur, Kku, Stan Shebs, Kimiko, Hike395, Kat, Pollinator, Robbot, Auric, MPF, Kerttie, MingMecca, Guanaco, RcktScientistX, Lulubutterfingers, Trevalyx, Thorwald, Seffer, Rich Farmbrough, Erauch, Jkh.gr, Still, Aitch Eye, Poweroid, Kazvorpal, Woohookitty, Jwanders, Rjwilmsi, WriterHound, RussBot, Gaius Cornelius, Dialectric, Ospalh, SmackBot, OrgulloKMoore, McGeddon, Cacuija, Squiddy, Kappus, Bluebot, Athanor~enwiki, Can't sleep, clown will eat me, Abrahami, Gobonobo, Calibas, Joseph Solis in Australia, Legaia, Meng.benjamin, Blissfulpain, JamesAM, Pastafarian Nights, AntiVandalBot, Shirt58, Tillman, Aquaponics, Dptalbot, Magioladitis, Xantres, Gennaroc, JaGa, Peter coxhead, MartinBot, Naniwako, Chiswick Chap, Lights, DrMicro, Qxz, Cerebellum, Doug, Phina.v, Kurasu, Kurihaya, Dymonite, MATThematical, Batpoep, Namazu-tron, Arjayay, Aaog, DumZiBoT, Addbot, Youre dreaming eh?, Glane23, Quercus solaris, Lightbot, Martin Hanson, Legobot, Yobot, Misssamerica, AnomieBOT, Apothecia, Anna Frodesiak, RibotBOT, Amaury, FrescoBot, KirbyRandolf, Pinethicket, Narfiol, EmausBot, Look2See1, Matthewcgirling, Erianna, ClueBot NG, PaleCloudedWhite, Alexander E Ross, Braincricket, Helpful Pixie Bot, Wbm1058, Northamerica1000, Hermes7979, Ma Hzi Wong, Jalaber, ChrisGualtieri, SFK2, Leugimap, PhantomTech, Tortie tude, HalfGig, Monkbot, Jordanhargravewiki and Anonymous: 52

- **Composting toilet** *Source:* https://en.wikipedia.org/wiki/Composting_toilet?oldid=749202251 *Contributors:* Quercusrobur, Kku, Julesd, Andrewman327, Wetman, Lumos3, Pfrishauf, Donreed, Kowey, KellyCoinGuy, Alan Liefting, Everyking, Pgan002, Alexf, Jm butler, Quadell, Onco p53, Ice Czar, Mrtrey99, Sam Hocevar, JTN, JesterXXV, Vsmith, Gjm, ESkog, CanisRufus, Vortexrealm, Batneil, Espoo, Carbon Caryatid, Cdc, Velella, Alai, Carbenium, Mindmatrix, Polyparadigm, Pol098, Rjwilmsi, Salix alba, SeanMack, Ewlyahoocom, Jrtayloriv, Mathrick, Wavelength, Jimp, Mukkakukaku, Carllindstrom, Splash, Gaius Cornelius, Nicke L, Irishguy, Diotti, Arthur Rubin, Jsheffield, Tevildo, GraemeL, Wakingdreaming, Carlosguitar, Luk, Rudy23, SmackBot, Reedy, KVDP, Hmains, Ladislav the Posthumous, Deli nk, Brinerustle, A. B., GerK, Chendy, OrphanBot, BobJones, Mitrius, Parrot of Doom, J.smith, Euchiasmus, Gobonobo, Ckatz, Rkmlai, TastyPoutine, Vonvon, M855GT, Peter Horn, Hu12, Natronomonas, Joseph Solis in Australia, IanOfNorwich, Tawkerbot2, ChrisCork, JForget, CmdrObot, Doctormatt, Gonefishingforgood, Phil in the 818, Webaware, Thijs!bot, Epbr123, Headbomb, MichaelMaggs, AntiVandalBot, Malvineous, Smartse, Lfstevens, MortimerCat, DuncanHill, Eurobas, SiobhanHansa, Magioladitis, Appraiser, 2share, Euhlig, ZackTheJack, Sustainableyes, WLU, Curtbeckmann, GTZ-44-ecosan, Heather hope, Transisto, FANSTARbot, Graham,kl, Knulclunk, Jorfer, Harmonyshenk, Signalhead, VolkovBot, Magnvss, JBazuzi, TXiKiBoT, DocteurCosmos, Broadbot, Temanning, Philip W Bush, Pixeljim, Agyle, PeterEasthope, Lamro, Doc James, Red58bill, Logan, Ponyo, Grahamken, Yousifm, TX55, Alatari, Tatterfly, Jon R W, Dforrester, ClueBot, Wikievil666, FrankyBoy1, Plastikspork, EoGuy, Drmies, Dylan620, Kafka21, 7&6=thirteen, Jcbrodie, XLinkBot, Delicious carbuncle, Jweaver214, Dthomsen8, Laughton.andrew, Addbot, Ashanda, MrOllie, C1614, Numbo3-bot, Hammerdownjonas, OlEnglish, EcoEthic, Ochib, Sanderlands, Yobot, Avenvonkrieger, AnomieBOT, Nutriveg, Six words, Killiondude, Greenthumb247, NickK, MauritsBot, Ecowaters, Gaialore, Biolet, Anna Frodesiak, GrouchoBot, Mregelsberger, Twirligig, BlazedAbyss, Ecopreneur, Misiekuk, PeterEastern, Leptosome, Mikal42, AstaBOTh15, Pinethicket, RDavey1314, Mmday, Cnwilliams, Julianplenti, Tbhotch, Mean as custard, Otutusaus, Dcirovic, DavidBeechey, Denver & Rio Grande, Noodleki, Hazard-Bot, Hyronimus299, ClueBot NG, Jphcz, Pbmaise, Riednelson, Nishaca, Helpful Pixie Bot, Mathew cloacina, George Ponderevo, Northamerica1000, Mark Arsten, Dexbot, DBeechey, Ecotoiletman, BradT123, Monadial, PeterLHarper, Stranman84, Kaitlinz, Robevans123, Mr. Smart LION, SkateTier, Cresyl, Filedelinkerbot, EvMsmile, EChastain, Krischan Makowka, Mll mitch, Gruster, Bio-CLC, Julielamberson, Cryacrem, JJMC89, Stewi101015, BOYLAN8888 and Anonymous: 183

- **Ecosynthesis** *Source:* https://en.wikipedia.org/wiki/Ecosynthesis?oldid=722354619 *Contributors:* Edward, Viriditas, Dialectric, Aeusoes1, Robofish, Gnoll110, ThisIsAce, Yakushima, Anna Frodesiak, Glacier2009, Helpful Pixie Bot, Northamerica1000, Johnsoniensis and Anonymous: 2

- **Ernie and Erica Wisner** *Source:* https://en.wikipedia.org/wiki/Ernie_and_Erica_Wisner?oldid=696104628 *Contributors:* Magioladitis, Duffbeerforme, AnomieBOT, GoingBatty, BG19bot, Lelindaelizabeth, OccultZone, Ephemeralcas and Ascii002

- **Folkewall** *Source:* https://en.wikipedia.org/wiki/Folkewall?oldid=738777057 *Contributors:* MBisanz, Velella, Zzuuzz, SmackBot, Thumperward, Basicdesign, Wafulz, Inclusivedisjunction, Red58bill, Malcolmxl5, Aitias, Pichpich, Wcoole, Anna Frodesiak, Look2See1, ClueBot NG, Wgolf, Lightweight Eddie, WhyWillSheNotDoTheNastyWithMe, GreenC bot and Anonymous: 8

- **Food Not Lawns** *Source:* https://en.wikipedia.org/wiki/Food_Not_Lawns?oldid=748717229 *Contributors:* SMcCandlish, Tryptofish, FrescoBot, Marcocapelle, HeatherJoFlores and Anonymous: 1

- **Grassed waterway** *Source:* https://en.wikipedia.org/wiki/Grassed_waterway?oldid=749259874 *Contributors:* Oddharmonic, BlueJaeger, Wavelength, RussBot, Dialectric, Bluezy, SmackBot, Cydebot, T@nn, NAHID, Addbot, Yobot, AnomieBOT, Anna Frodesiak, Evrardo, Look2See1, SporkBot, ClueBot NG, Widr, Antiqueight, Arr4, Cyberbot II, Kwaldroup2002, GreenC bot and Anonymous: 4

- **Holzer Permaculture** *Source:* https://en.wikipedia.org/wiki/Holzer_Permaculture?oldid=751721611 *Contributors:* Velella, Firsfron, Monkofthetrueschool, Malcolma, KVDP, Byelf2007, CmdrObot, Pdcook, Ilyushka88, WereSpielChequers, Cr0, The Bushranger, AnomieBOT, Anna Frodesiak, Glacier2009, BoundaryRider, BG19bot, Jeffersonfranklin, Bender the Bot and Anonymous: 9

- **Intercropping** *Source:* https://en.wikipedia.org/wiki/Intercropping?oldid=752533114 *Contributors:* Dysprosia, Pollinator, Alan Liefting, Germen, Erauch, Vortexrealm, Maurreen, Kjkolb, Linmhall, Kazvorpal, Jwanders, Strait, Salix alba, Chobot, Bgwhite, Kummi, Gaius Cornelius, Nirvana2013, Lckesdonkey, Garion96, SmackBot, McGeddon, CRKingston, Eskimbot, Cacuija, Gilliam, Chris the speller, Darth Panda, Jhml, KP Botany, Smartse, Steven Walling, Wassupwestcoast, Naohiro19, Mbuckingham, Markisgreen, Mikemoral, Der Golem, Mild Bill Hiccup, Tanketz, EmmaRubu, Kembangraps, Addbot, Tassedethe, BlazerKnight, Dyorkey, Materialscientist, Apothecia, Anna Frodesiak, Craig Pemberton, Ezhuttukari, Katach, Kibi78704, RjwilmsiBot, Skamecrazy123, Look2See1, Slightsmile, Dcirovic, Anir1uph, Matthewcgirling, Jsayre64, علم‌رو بن ڵلﺗﻮم, ClueBot NG, Minerv, Wbm1058, Gob Lofa, Northamerica1000, Tom Pippens, Jalaber, SMARTY 123, अनुनाद सिंह, YiFei-Bot, Jenjhall, TheEditor867, Weopi, Utters11, Bender the Bot, Carsonac, Zcarstvnz and Anonymous: 28

- **Keyline design** *Source:* https://en.wikipedia.org/wiki/Keyline_design?oldid=750182728 *Contributors:* Francs2000, Rich Farmbrough, Max Terry, Bender235, Thu, GrantNeufeld, Salix alba, Dialectric, CQ, SmackBot, RDBrown, Sholto Maud, Brimba, Radagast83, JSchinnerer, Yeomansplowchris1, HisSpaceResearch, FlyingToaster, TAMilo, Alphachimpbot, Erkan Yilmaz, Gueneverey, Piperh, Sfan00 IMG, Yamakiri, Highdozen, Cr0, LilHelpa, Anna Frodesiak, Miyagawa, Glacier2009, Allthingstoallpeople, Stephen.J.Arnold, Deepbiosoil, Cockaroach, Helpful Pixie Bot, Gob Lofa, Cyberbot II, Me, Myself, and I are Here, Rodquiros, Equinox, CarbonCyclist and Anonymous: 14

- **Leaf mold** *Source:* https://en.wikipedia.org/wiki/Leaf_mold?oldid=744682462 *Contributors:* Bryan Derksen, Stephen Gilbert, Rgamble, Quercusrobur, Infrogmation, Stan Shebs, Glenn, Jose Ramos, Sheridan, Guettarda, Enric Naval, Vortexrealm, Paleorthid, Velella, Jwanders, Wavelength, Splash, Welsh, JezzBrookes, BorgQueen, SmackBot, Pgk, Cacuija, Shai-kun, Squiddy, JohnCub, Lucy Cassidy, Croton, Soulbot, J A Ellam, Jeepday, Fences and windows, Red58bill, SieBot, FerdinandFrog, Lokionly, Addbot, Lightbot, Yobot, Apothecia, Anna Frodesiak, Wikininja2.0, Finalius, Look2See1, Sad whale and Anonymous: 18

- **Vegan organic gardening** *Source:* https://en.wikipedia.org/wiki/Vegan_organic_gardening?oldid=739129849 *Contributors:* Quercusrobur, Ronz, Steinsky, Pollinator, Gentgeen, Pengo, Alan Liefting, Bender235, Mwanner, Viriditas, Maurreen, Pearle, Paleorthid, Velella, RJFJR, Bobrayner, Mel Etitis, ADeveria, Rjwilmsi, Salix alba, Fel64, Ewlyahoocom, Gaius Cornelius, Jrideout, Dialectric, Nirvana2013, IreverentReverend, That Guy, From That Show!, SmackBot, Amcbride, Bluebot, Nomenclator, Andrew c, Mksword, Iridescent, Ayanoa, Luna Santin, Shousokutsuu, Michig, Think outside the box, Sustainableyes, Michaelbedar, Greenwoodtree, Jon-mikel, Idioma-bot, Joopboer, Barkeep, MaynardClark, Randy Kryn, Dakinijones, Excirial, XLinkBot, Flimflam2008, Verbal, Pinus jeffreyi, Yobot, AnomieBOT, LilHelpa, Betty Logan, Anna Frodesiak, MikeYoungRealEstate, Legion23, Veggieburgerfan, The Freegan Vegan, Jay-Sebastos, Hineta, ClueBot NG, PaleClouded-White, Helpful Pixie Bot, Northamerica1000, Rowan Adams, Antifarmer, Lose2fat, Zylacourtney and Anonymous: 28

- **Village Community Co-operative** *Source:* https://en.wikipedia.org/wiki/Village_Community_Co-operative?oldid=749776538 *Contributors:* Bearcat, Grahamec, Niceguyedc, AnomieBOT, E-historyprospect, DaltonCastle and Atlantic306

- **Village Homes** *Source:* https://en.wikipedia.org/wiki/Village_Homes?oldid=749500317 *Contributors:* Jengod, Carlossuarez46, Klemen Kocjancic, Grutness, SmackBot, Fgrammen, Mr3641, Theleek, Biruitorul, Barek, Afrothetics, FrescoBot, Twastvedt, Marcocapelle, InternetArchiveBot, GreenC bot and Anonymous: 4

- **Waru Waru** *Source:* https://en.wikipedia.org/wiki/Waru_Waru?oldid=742637581 *Contributors:* Alan Liefting, Wtmitchell, Kralizec!, Dialectric, Kevlar67, Themightyquill, CaTi0604, Silver seren, The Anomebot2, Oceanflynn, Smsarmad, Stone Violin, Dthomsen8, Jean.vivien.maurice, Addbot, AnomieBOT, Anna Frodesiak, Louperibot, EmausBot, Look2See1, Dcirovic, ClueBot NG and Anonymous: 11

- **Paul Yeboah** *Source:* https://en.wikipedia.org/wiki/Paul_Yeboah?oldid=746463866 *Contributors:* Magioladitis, Yobot, BG19bot, Celestinesucess, Adjoajo and Anonymous: 2

- **List of permaculture projects** *Source:* https://en.wikipedia.org/wiki/List_of_permaculture_projects?oldid=744327449 *Contributors:* Edward, Jengod, Alexf, Rich Farmbrough, Vsmith, Dejitarob, Shafaki, Velella, BD2412, Bgwhite, Jimp, Leighblackall, Nirvana2013, Number 57, SmackBot, Thumperward, Dandelion1, TerryE, CoulterTM, Anammasrur, Jac16888, Kanejamison, Emeraldcityserendipity, .alyn.post., Kshea19, Keith D, Terrek, OniOid, DASonnenfeld, VARGUX, Dusti, Jack Merridew, Jojalozzo, Lightmouse, Ryansb90, JL-Bot, Sfan00 IMG, Hippo99, EoGuy, Arjayay, Iohannes Animosus, XLinkBot, Dthomsen8, Lotusdog, AnomieBOT, Anna Frodesiak, Ecopreneur, FrescoBot, Glacier2009, Bulgar khan, Permaship, Hstory, Finca el zopilote, Mean as custard, John of Reading, Look2See1, Dewritech, Rafaele Joudry, Cockaroach, Permamax, Foxyfire, Mettaexperiment, Permacultureguru, Widr, Antiqueight, Mouramoor, Philospelunk, BG19bot, Christianshearer, Iamharb, BattyBot, Bruce W. Williams, Bluth12, StephCrmpi, Mogism, P3Permaculture, Cevansuk, Chrysbisson, Ruifcnunes, Kosmic roots, Jc.nicaise, Sbaxendell, Permie girl, BarefootandBroke, Suavicm, Chrismonk3, WWOOF Thailand, Hstory2000, Breedentials, Narky Blert, Enzo at Permaculture Education, PermaRay, Rubbish computer, Nx7000, Brucebebe and Anonymous: 80

31.9.2   Images

- **File:2009NativeAmericanRev.jpg** *Source:* https://upload.wikimedia.org/wikipedia/commons/d/d7/2009NativeAmericanRev.jpg *License:* Public domain *Contributors:* United States Mint Historical Image Library *Original artist:* United States Mint

- **File:2010_WiseWords_community_garden_NewOrleans_4855100893.jpg** *Source:* https://upload.wikimedia.org/wikipedia/commons/1/19/2010_WiseWords_community_garden_NewOrleans_4855100893.jpg *License:* CC BY 2.0 *Contributors:* Beds *Original artist:* Bart Everson from New Orleans, Louisiana, USA

- **File:Aegopodium_podagraria1_ies.jpg** *Source:* https://upload.wikimedia.org/wikipedia/commons/b/bf/Aegopodium_podagraria1_ies.jpg *License:* CC-BY-SA-3.0 *Contributors:* Own work *Original artist:* Frank Vincentz

- **File:Amanita_praecox_86186.jpg** *Source:* https://upload.wikimedia.org/wikipedia/commons/d/de/Amanita_praecox_86186.jpg *License:* CC BY-SA 3.0 *Contributors:* This image is Image Number 86186 at Mushroom Observer, a source for mycological images. *Original artist:* This image was created by user Dan Molter (shroomydan) at Mushroom Observer, a source for mycological images.

- **File:Ambox_globe_content.svg** *Source:* https://upload.wikimedia.org/wikipedia/commons/b/bd/Ambox_globe_content.svg *License:* Public domain *Contributors:* Own work, using File:Information icon3.svg and File:Earth clip art.svg *Original artist:* penubag

- **File:Ambox_important.svg** *Source:* https://upload.wikimedia.org/wikipedia/commons/b/b4/Ambox_important.svg *License:* Public domain *Contributors:* Own work, based off of Image:Ambox scales.svg *Original artist:* Dsmurat (talk · contribs)

- **File:Arborloo-en.svg** *Source:* https://upload.wikimedia.org/wikipedia/commons/a/a2/Arborloo-en.svg *License:* GFDL *Contributors:* Own work Inspired by Peter Morgan, *Toilets that make compost: Low-cost, sanitary toilets that produce valuable compost for crops in an African context*, EcoSanRes, Stockholm Environment Institute, 2007, ISBN 978-9-197-60222-8, figure 2.1, page 3 (read the book online) *Original artist:* SuperManu

- **File:Arborloo_(5567538554).jpg** *Source:* https://upload.wikimedia.org/wikipedia/commons/3/3f/Arborloo_%285567538554%29.jpg *License:* CC BY 2.0 *Contributors:* https://www.flickr.com/photos/gtzecosan/5567538554/ *Original artist:* SuSanA Secretariat

- **File:Armillaria_mellea_620.jpg** *Source:* https://upload.wikimedia.org/wikipedia/commons/e/ee/Armillaria_mellea_620.jpg *License:* CC BY-SA 3.0 *Contributors:* This image is Image Number 620 at Mushroom Observer, a source for mycological images. *Original artist:* This image was created by user Nathan Wilson (nathan) at Mushroom Observer, a source for mycological images.

- **File:Artocarpus_heterophyllus_fruits_at_tree.jpg** *Source:* https://upload.wikimedia.org/wikipedia/commons/d/d0/Artocarpus_heterophyllus_fruits_at_tree.jpg *License:* CC-BY-SA-3.0 *Contributors:* Own work *Original artist:* Photo taken by User:Ahoerstemeier on June 28, 2003 in Chaiya, Surat Thani Province, Thailand.

- **File:Bedrijfsafval.jpg** *Source:* https://upload.wikimedia.org/wikipedia/commons/3/3a/Bedrijfsafval.jpg *License:* Public domain *Contributors:* No machine-readable source provided. Own work assumed (based on copyright claims). *Original artist:* No machine-readable author provided. Fun4life.nl assumed (based on copyright claims).

31.9.3   Content license